EATING IN THE DARK

EATING IN THE DARK

*America's Experiment with
Genetically Engineered Food*

KATHLEEN HART

Pantheon Books, New York

Pantheon Books and colophon are registered
trademarks of Random House, Inc.

Library of Congress Cataloging-in-Publication Data

Hart, Kathleen.
Eating in the dark : America's experiment with
genetically engineered food / Kathleen Hart.
p. cm.
Includes index.
ISBN 0-375-42070-3
1. Genetically modified foods—United States.   I. Title.
TP248.65.F66 H37 2002   363.19′29—dc21   2001058842

www.pantheonbooks.com

Book design by M. Kristen Bearse

Printed in the United States of America
First Edition
2   4   6   8   9   7   5   3   1

In memory of my father,

Robert Hart

CONTENTS

EATING IN THE DARK

THE FOOD EXPERIMENT

On October 16, 2000, Grace Booth, the residential director of a home for troubled youth in the San Francisco Bay area, made enchiladas for lunch. She used basic ingredients—garlic, cheese, chicken, enchilada sauce, and corn tortillas. Ten minutes after eating the meal, her arms, head, and back started itching intensely. Her lips became tingly and swollen. Her throat and airway were closing, and when she tried to call for help from co-workers, her voice would not come. She was rushed to the hospital in an ambulance, where she went into anaphylactic shock and almost died.

Grace was one of dozens of Americans who reported having a serious reaction to foods containing corn in the summer and fall of 2000, around the same time a variety of genetically engineered corn called StarLink was first discovered in taco shells. StarLink was engineered by the biotechnology company Aventis to produce a toxic protein that kills insects—a built-in pesticide that would save farmers the trouble and expense of spraying their crops. The corn was supposed to be used only for animal feed. The Environmental Protection Agency had banned StarLink from the food supply in May 1998 because the toxic protein in the corn is nearly indigestible and some allergy experts fear it may trigger allergic reactions in susceptible people.

Nine months after her brush with death, on July 17, 2001, Grace traveled to Arlington, Virginia, to speak before a panel of world-renowned allergy experts. The EPA convened the panel of doctors and scientists to assess the health risk StarLink corn poses to the American public. While there was no scientific evidence showing that StarLink corn had sickened Booth or others, the very possibility that an unapproved genetically engineered food on the market might cause allergic reactions troubled some physicians and unnerved many consumers. Dr. Dean Metcalfe, a preeminent allergist at the National Institutes of Health in Bethesda, Maryland, and a member of the EPA advisory panel, called the release of genetically engineered StarLink corn into the human food chain "a massive, unintended, natural experiment" taking place in the United States.

At the conclusion of the EPA's meeting, another panel member, Dr. Marc Rothenberg, a pediatrician at Children's Hospital Medical Center in Cincinnati, said, "I believe that the StarLink is potentially just the tip of the iceberg in terms of future public concerns" that are likely to arise from genetically modified foods—foods that, unlike StarLink corn, the government has allowed to enter the food supply legally. In fact, the illegal StarLink variety made up only a tiny fraction of the millions of acres of genetically engineered corn and soybeans that had been grown across the American farm belt since 1996. Five years from now, Dr. Rothenberg warned, an entirely different and unexpected problem might be found with another genetically modified food product.

Many of our most familiar foods are now made from strange new ingredients: genetically modified organisms, or GMOs. You can't see them. You can't taste them. And they're not listed on food labels in America. The Food and Drug Administration says genetically modified corn and soybeans are as safe and nutritious as traditional corn and soybeans, but 350 million Europeans and 125 million Japanese refuse to eat them. They are worried about possible latent health effects or environmental problems from these new biotech crops.

It's hard to get through a day without eating soybeans in one form or another. The versatile bean can be found in about two-thirds of the processed foods on our grocery store shelves. Food manufacturers use soy oil, lecithin, meal, protein, and grits in thousands of common food items, from bread, mayonnaise, and canned tuna to ice cream, cookies, and frozen yogurt bars, and even dog food. A bushel of soybeans yields about 48 pounds of protein-rich meal and 11 pounds of soy oil. Most of the soybean meal is fed to animals, but soy oil is the top edible oil used in the United States. Soybean proteins are used in bakery goods, confections, meat products, and nutritional supplements. Soy protein concentrate is the main ingredient in the soy burgers that have become popular with health food enthusiasts and vegetarians in recent years.

Beginning with the 1996 harvest, American parents began feeding their babies bottles of formula made from genetically engineered soybeans. Few were aware, however, that the formula contained genetically engineered soybeans. These patented soybeans were created in laboratories of the chemical company Monsanto. They contain genes from bacteria, viruses, and petunias. The magic of Monsanto's patented beans lies in their ability to survive dousing with the company's top-selling weed-killer, Roundup, which makes weed management easier. Roundup Ready soybean plants soak up as much Roundup as farmers wish to spray on them, while weeds growing in the field near the soybeans, lacking the patented mix of genes, wither and die.

What began as a trickle of gene-altered soybeans into America's food stream in 1996 would grow to a torrent by 2001. U.S. farmers planted one million acres of Roundup Ready soybeans in 1996. They were harvested in the fall and mixed in with regular soybeans. In 1997 farmers grew Roundup Ready soybeans on 12 million acres, accounting for 17 percent of the soybean crop. In 1998, 38 percent of the nation's soybean harvest was genetically engineered, climbing to more than 50 percent in 1999. By 2001 more than 60 percent of the soybeans planted in America were Roundup Ready.

Every day millions of Americans consume breakfast cereals, fruit drinks, chips, and a host of common snack foods containing a new kind of patented corn that makes its own pesticide. The pesticide in the corn comes from the genes of the soil bacterium *Bacillus thuringiensis,* or Bt for short. The corn plant's roots, stalk, leaves, pollen, cob, and kernels all produce doses of a genetically engineered toxin. To get corn plants to produce this pesticide, scientists shoot virus and bacteria genes into corn embryos, hit or miss, until the foreign genes "take" in one of the embryos, which grows into a modified corn plant.

Some of the genetically engineered corn in the foods now on our dinner plates comes from hybrids that make their own pesticide *and* withstand spraying with weed-killers.

The EPA licensed the sale of the first Bt corn variety in 1995. All of the genetically engineered corn grown in the United States, with the exception of StarLink, was approved for human food. Farmers first planted genetically modified corn seed in 1996. In 1997 U.S. farmers harvested about four million acres of Bt corn, accounting for 5 percent of the eighty-million-acre domestic crop. Genetically modified corn, engineered to tolerate weed-killers or produce Bt toxins, accounted for 25 percent of the 1998 corn harvest. In 1999 genetically engineered corn made up about 35 percent of the domestic crop, dropping off to about 25 percent during the 2000 and 2001 growing seasons.

Corn is America's largest and oldest food crop. About 80 percent of the annual corn harvest goes to animal feed. The remaining 20 percent, about two billion bushels, is used for human food or for ethanol. Corn is a mainstay of the food industry, with corn syrup, cornstarch, corn gluten, high-fructose corn syrup, and cornmeal going into a significant proportion of processed foods. The soft drink market alone uses more than 500 million bushels of corn a year. Breakfast cereals and snack foods account for another 140 million bushels.

The genetic alteration of two of America's staple food crops took place quietly, without the awareness of most U.S. citizens. In February 1999 two-thirds of American adults responding to a national survey did not know that supermarkets were selling genetically modified foods. Only 3 percent of Americans knew they were eating food made with genetically engineered soybeans; only 6 percent knew they were eating genetically engineered corn.

The FDA has assured the American public that genetically modified foods are safe and will not present long-term human health problems. To this assertion, Dr. Martha Herbert, a pediatric neurologist at Massachusetts General Hospital and Harvard Medical School, responds, "There's just not enough data for officials to have made a strong statement about the safety of genetically modified foods."

The FDA's position is based on scientific knowledge that was available to researchers in the early 1990s, when genetically engineered corn, soybeans, potatoes, and canola were created. In company-sponsored tests, Roundup Ready soybeans were added to the diets of rats for 28 days, chickens for 6 weeks, cows for 4 weeks, catfish for 10 weeks, and bobwhite quail for 5 days. No adverse health effects were found. The FDA did not require Monsanto or other biotechnology companies to perform feeding studies of their new gene-altered crops on primates or humans.

In effect, critics argue, since the 1996 harvest the entire U.S. population has been part of an uncontrolled experiment to demonstrate the long-term safety of gene-altered corn and soybeans. Yet no one has signed a consent form.

From the standpoint of science, the experiment is poorly designed. There are no control groups. Scientists are not measuring the "doses" of gene-altered foods that various subgroups of the population are receiving. Doctors are not monitoring the population for unexpected health effects that might be linked to ingestion of particular genetically modified foods.

As a business strategy, the experiment is both clever and daring. Because the foods were never labeled, consumers who suffer any unanticipated health effects will have no way of proving that those effects are attributable to one particular genetically engineered food or another. It will be hard for any injured parties to assign blame or determine liability.

From a public health perspective, the experiment is a nightmare. Dr. Herbert says that without food labeling it is virtually impossible to do public health monitoring. "The food is already widely distributed with no traceability," she observes. "One day the store may stock a genetically modified product and the next day not." She believes it is possible that allergies and other "disease endpoints" could result from eating these new gene-altered foods.

Eating in the Dark tells the story of how the Food and Drug Administration and other federal agencies helped the biotechnology industry to impose gene-altered food products on an unwitting American pub-

lic and an unwilling international community. I first learned about
genetically engineered corn and soybeans in May 1997, during an interview with the representative of a British trade group. In the fall of
1996, when U.S. grain exporters started mixing genetically engineered
soybeans in with shipments of regular soybeans, the British Retail
Consortium had warned of a possible boycott unless the practice was
ended. Rupert Hodges, a spokesman for the consortium, said that
British grocery store owners were growing frustrated because no one
in America seemed willing to seriously consider British consumers'
worries about genetically modified soybeans and corn. Citizens in
every corner of the European Union were pressuring grocery store
owners to label clearly any food containing genetically modified organisms. Fifteen food chains in the United Kingdom, France, Denmark, Switzerland, Germany, and Sweden had just signed an open
letter to the "U.S. commodity agribusiness," dated May 28, 1997, entreating growers and shippers to keep genetically modified corn and
soybeans separate from conventional corn and soybeans.

I called Monsanto, whose Roundup Ready soybeans were causing
such a stir, to find out if the company would agree to separate and
label genetically modified beans, as the Europeans were asking. A
Monsanto spokeswoman told me that it would be impossible to segregate Roundup Ready soybeans from other soybeans. Throughout the
corn belt, gene-altered beans and conventional beans are harvested
using the same combines, stored in the same grain elevators, moved
on the same conveyor belts, trucks, and barges, and processed together for livestock feed and food, she explained.

What about the idea of giving customers a choice? She gave me
what sounded like a carefully rehearsed answer: "They look alike,
they taste alike, they function alike and they've been approved around
the world as a safe, new product."

Next I called the U.S. Department of Agriculture, where an assistant to Agriculture Secretary Dan Glickman assured me that genetically engineered soybeans and corn are nutritionally the same as
regular soybeans and corn. The Europeans are not basing their objections on science, he said. "It's based on emotion."

When I called the FDA to find out the government's position on
genetically modified organisms, a spokesman told me there's "no sub-

stantial difference between genetically engineered foods and their conventional counterparts." I was puzzled. "If it's exactly the same, why do the Europeans have a problem with it?" I asked. "It's a question of public acceptance over there," the FDA spokesman responded.

The vaguely dismissive responses of these USDA and FDA officials left me wondering just what the differences *are* between genetically engineered and traditional foods and precisely why European citizens have shunned them.

The FDA did not plan the gene-altered food experiment that began in North American households following the 1996 harvest, but the agency helped the biotechnology industry launch it. Funded by taxpayers and charged with safeguarding America's food supply, FDA officials have steadfastly maintained that patented food plants injected with genes from viruses, bacteria, and other organisms are no different from natural crops. By waiving the requirement for expensive safety testing of genetically modified foods—the kind of testing required for new drugs and food additives—the FDA has put corporate interests ahead of public welfare.

In refusing repeated consumer pleas for labels on genetically engineered foods, the FDA has prevented Americans from making an informed choice about the foods we eat and feed our families. Around the world, from Europe to Asia, doctors and ordinary consumers are watching us, with curiosity, to see whether the genetically modified foods in our diet will turn out to be safe. Especially, they are watching our children, the first generation to be raised on these new biotech foods.

ONE

FORCE-FED CONSUMERS

In the early morning hours of Sunday, September 28, 1997, a small band of citizens crept into a field in County Carlow, fifty miles south of Dublin. Illuminated by the glow of yellow lights from a nearby industrial plant, they set about slashing and digging up every genetically modified sugar beet that had been planted by Monsanto on the one-acre plot. The following morning an anonymous caller from a previously unknown group identifying itself as the Gaelic Earth Liberation Front left a message with the Genetic Engineering Network in London. The caller said, "This was the first genetically engineered crop in Ireland, and hopefully it will be the last."

Word of the sabotage against Monsanto's experimental crop spread quickly across Ireland. Patricia McKenna, Ireland's Green Party member of the European Parliament, applauded the activists, telling *The Irish Times* that they deserved "full praise." Officials at the government-sponsored research site in Carlow condemned the vandalism. The nation's relatively young biotechnology organization, the Irish BioIndustry Association (IBIA), also deplored the act and noted that it is not the sort of thing that happens very often in Ireland.

One of the "Carlow diggers," as the activists became known, later said the Gaelic Earth Liberation Front was "not a group with a constituted membership." Rather, several concerned citizens had come together somewhat spontaneously to rid the countryside of mutant plants that they believed posed an immediate and unacceptable threat to their future security. "At the end of the action there was no sudden sense of great achievement and no real celebrations, instead there was something simpler, a feeling that we came, we dug and we made Ireland GE free again," a self-proclaimed participant said in a letter posted on the Internet. "We all believe that a future untainted by runaway biological pollution is a future worth fighting for. This isn't terrorism, it's realism."

Many law-abiding Irish consumers supported the Carlow action. Five months earlier, on May 1, 1997, the Irish Environmental Protection Agency had granted Monsanto the first license in Ireland for a deliberate release of genetically modified organisms into the environment. Theretofore research on genetically engineered bacteria and plants had been performed indoors, under tightly controlled conditions, to prevent birds, bees, butterflies, and breezes from spreading pollen and seeds produced by gene-altered crops into the countryside.

Clare Watson had started the anti–genetic engineering group Genetic Concern! in April 1997 to fight Monsanto's proposed outdoor field experiment through legal channels. She immediately sought a high court review of the Irish EPA's decision. The court granted her request for a review but declined to issue an injunction preventing Monsanto from planting its sugar beets, which were genetically engineered to resist dousing by the company's weed-killer Roundup. Monsanto put its experimental sugar beet crop into the ground on May 27, 1997. The high court's legal review of the EPA's license decision was not scheduled to begin until December 10, 1997, by which time the genetically modified sugar beets would have been harvested. Sugar is a biennial crop and doesn't generally flower in its first year, although it does occasionally happen (when it is called "bolting"). This was the risk that worried Genetic Concern!

"Why are the Irish so passionate about this issue?" I later asked Clare Watson and Quentin Gargan, a spokesperson for Genetic Concern! and owner of a health food business.

"The public is frightened by the complexity of the issue," answered Watson, who grew up on an organic farm in Cork. "The issue of control over the food supply, too, grows out of the history of the potato famine. Who's controlling our agriculture? Our history tells us that who controls the food supply is a life and death matter," she explained. Foreign ownership of agriculture is a subject that touches an emotional chord with the Irish. While 1.5 million people died of starvation or diseases caused by hunger and poverty in Ireland during the potato famine between 1845 and 1850, barley and oats were being shipped out of the country to England.

"Ireland regards itself as a country of tourism, and its green image has to be real," added Gargan. The Irish boast of having the purest water in the European Union—and some of the most beautiful landscapes. Genetic engineering would encourage large tracts of monoculture sugar beets, which in turn could transform the look of the countryside and the social fabric of small farming communities.

"Also related is the need for genetic diversity," Clare said, "and Monsanto's intensive agriculture—promoting one strain—is opposed to that." Rather than genetically engineering artificial strains and having just a few varieties available for each crop, as the multinational chemical corporations that have been acquiring seed companies are doing, she believes future food security depends on preserving and improving the hundreds of varieties that already exist. "We should be going back and looking at what we had. But there's no profit for the multinational companies in just allowing farmers to go back to the old varieties."

Gargan, whose company, Wholefoods Wholesale, provides health food stores with tofu and other natural products, described his initial reaction to the arrival of Monsanto's gene-altered soybeans in Ireland. "We went berserk when we found out about the genetically modified soya mixed in with the soybean shipments coming from the States!" he exclaimed. His company responded by finding soymilk and tofu suppliers who would guarantee GM-free sources of soybeans.

"We're getting raisin producers in California to use coconut oil instead of soy oil to coat the raisins. The same with fruits from Australian growers," Gargan continued. "Monsanto made a grave mistake in choosing to genetically modify soya first, because it's the icon of the food industry."

"What is your greatest personal concern about genetically engineered foods?" I asked them.

"You wait for another thalidomide, or something terrible to go wrong," Gargan answered. "I think the allergies problem is real.

There will be no time to trace the cause back if people start developing allergies. We don't have a database on who's allergic to petunia."

"How do you know, in the genetic manipulation, you're not displacing a gene that's supposed to be expressed?" Clare asked. She paused for a moment, then added, "The problem with all of this unlabeled genetically modified food is that if something goes wrong, it goes wrong badly."

In November 1997 I traveled to England to learn more about the millions of acres of genetically modified crops growing in the United States. I was eager to talk to scientists, shopkeepers, and regular consumers to find out firsthand why they opposed gene-altered crops and the foods made from them. I came from a country where farmers had just harvested millions of bushels of gene-altered corn and soybeans, yet very few of the people who would eat the foods had any idea they were genetically modified. U.S. food industry leaders seemed unfazed by the notion of corn engineered to make its own pesticide and soybeans designed to soak up chemical weed-killers. But the insistence of American bureaucrats that gene-altered crops are essentially the same as normal crops seemed to gloss over what might be important biological differences.

My first stop was Greenpeace, the nerve center of a nascent movement to keep the United Kingdom—and Europe—free of genetically engineered plants and foods. The London office of the international environmental group is housed in a large, airy building in Islington, a chic village in the north of the city. I met with Douglas Parr, one of the chief UK campaigners against genetically modified food. A soft-spoken man with a serious demeanor, Parr displayed a solid command of the science issues that lie at the heart of the public's worries about genetic engineering. He gave me a report he had just published, entitled "Genetic Engineering: Too Good to Go Wrong?" In it he laid out, in spare yet chilling scientific detail, twelve case studies of genetic engineering experiments that have gone terribly wrong.

The main problem with genetically engineered plants, animals, and microorganisms, Parr found, is their unpredictability. They are not stable, and often the inserted genes do not behave as expected. In one experiment to change the coloration of petunias, scientists genetically engineered the flowers to change their color from white to salmon pink. Quite unexpectedly, several weeks into the growing season, a majority of the petunias lost their salmon color and returned to white—a phenomenon the scientists chalked up to a three-week period of extremely high temperatures. In another case, a single gene that enhances the production of growth hormone was inserted into various farm animals, with hopes of making them grow faster. The inserted gene gave sheep diabetes and made them die young. A genetically engineered ram failed to mature sexually. In pigs the elevated levels of growth hormone caused arthritis, stomach ulcers, and heart and skin abnormalities. These and other examples showed that genes are not constant and fixed attributes of living things; they interact with other genes in the organism and are affected by their environment.

"There's a lot of unease about genetically engineered food here, and a great deal of concern about food integrity in general. The public has a distrust of science and a distrust of food authorities because of mad cow disease," Parr told me. A devastating epidemic of mad cow disease swept across the UK in the 1980s and early 1990s. The gruesome disease, called bovine spongiform encephalopathy, or BSE, leaves holes in the brains of infected cows. The epidemic led to the slaughter of hundreds of thousands of cattle.

From the first diagnosis of BSE in cows in 1986, British officials assured the public that there was no evidence to suggest the disease could spread to humans. In 1996 authorities finally acknowledged that ten deaths in people under the age of forty-two from a deadly brain disease were likely linked to exposure to BSE in beef. The individuals were diagnosed with a variant strain of Creutzfeldt-Jakob disease (CJD), a rare syndrome that ordinarily strikes people in middle age.

Creutzfeldt-Jakob disease is a horrifying, rapidly progressive, fatal disease with no treatment and no cure. Victims often suffer vision problems, jerky movements, and dementia before they die. CJD has a long incubation period between exposure to the infective agent and appearance of symptoms—from eighteen months to as many as twenty-five years, scientists believe. By the end of the century, dozens

more Britons were diagnosed with variant CJD. In May 2001 the
number of cases in the UK reached one hundred, and no one knows
how many victims will eventually succumb to the disease.

Scientists now believe the British BSE epidemic resulted from
changes in animal feed production methods: the feed industry began
to process rendered brains and carcasses of cows and sheep into
ground meal and then add the meal to cattle feed. Many cows became
infected before the source of contamination was discovered.

Through this long, painful experience with mad cow disease, the
British public has developed a great deal of mistrust about the pro-
nouncements of scientists and government officials who assert that
"there is no evidence of harm" from a new food product or an unproven food production method.

Parr believes that the risks from genetically modified organisms
mirror those of mad cow disease. In each case the food industry embarked on a new food production technique before its ramifications
for public health were entirely understood. In each case, scientists argued that because they had not found any problems with the food, it
was safe for humans to eat.

"People are upset because they see no need for this new genetically
engineered food. It's being pushed ahead blindly by the government,
yet the hazards are potentially colossal, and irreversible," Parr said.
"Once these organisms are released to the environment, they most
likely will be uncontainable."

Like the Carlow activists, some British citizens were worried that
by allowing biotech companies to test their experimental crops outside, the government was taking unacceptable risks with their food
supply and environment. In the fall of 1997, anonymous citizens in
the UK had pulled up a field of gene-altered potatoes and a field of engineered oilseed rape. Parr was quick to disassociate Greenpeace from
those actions. In fact, a large number of grassroots anti–genetic engineering networks were springing up around the country.

"Do you think you will be able to stop genetically engineered crops
from being grown in Great Britain?" I asked. "And do you think you
can stop the import of genetically modified foods from the United
States?"

On that afternoon Parr did not seem optimistic about the chances of keeping the United Kingdom free of genetically modified organisms. "If it were a Europe-only issue, it could possibly be stopped, but there's the U.S. You've got a huge institutional lineup for it—the biotechnology companies, the government, the food industry, the farmers," he replied. "There's a juggernaut."

He paused a moment, then concluded, "It's a difficult campaign, because there are no bodies in the streets to point to with genetic engineering. It's more the feeling of uneasiness, a worry about the unknown risks."

I considered Parr's use of the word "juggernaut" to describe the onslaught of agricultural biotechnology. The institutions, money, and power lined up behind agricultural biotechnology in the United States do indeed seem unstoppable. For more than two decades, pharmaceutical and agricultural chemical companies have been promoting biotechnology as a way to speed the development of new drugs and farm chemicals, as well as to increase their profitability. Just as physicists rocked the world in the mid-1940s when they split the atom and unleashed its awesome destructive force, biologists probing the inner workings of genes would have a chance, three decades later, to demonstrate their power to change the shape of life itself. In 1973, when two California scientists inserted a gene from a toad into bacteria—creating a novel form of life—biotechnology became a potent new force in America. By the early 1980s business leaders from a broad spectrum of industries had become convinced that biotechnology would open new doors to future inventions and earnings.

Food processing companies were among the earliest investors in biotechnology research. In 1982 Campbell Soup Company contracted with DNA Plant Technology Corporation to conduct research on ways to improve the solids content of tomatoes. Heinz enlisted the Atlantic Richfield Company's Plant Cell Research Institute for similar research. Because every one-percent increase in the solids content of tomatoes was estimated at the time to be worth $80 million a year in savings on processing costs, both companies hoped to realize hefty returns on their research investments.

In the mid-1980s American Home Products conducted research to see if popcorn plants could be engineered with built-in salt-and-butter flavor. Hershey Foods tried to develop new varieties of cocoa beans that were less bitter than regular cocoa. General Foods funded research to develop a low-caffeine coffee bean plant. Minute Maid and Coca-Cola officials were interested in citrus research carried out by a USDA plant geneticist who made orange juice from cultures of orange plant cells in test tubes. In the late 1980s the Chocolate Manufacturers Association, Quaker Oats, and the Popcorn Institute funded various plant biotechnology research projects at Purdue University.

"Biotechnology is, without a doubt, one of the most precocious of discoveries, quickly moving into the center of biological sciences," Theodore Hullar, chancellor of the University of California at Davis, declared in 1993. At the core of biotechnology "is control of genetic codes, of restructuring the genome, of making new forms of life, of tailoring life as we would have it be."

For years the National Academy of Sciences and the nation's top agriculture research universities have zealously pushed the idea that the twenty-first century will be based on a biotech economy. They envision a future where all our food, feed, fiber, medicines, energy, and even industrial chemicals come from bioengineered plants and animals. "The plant revolution is proceeding. It is being propelled by the skillful applications of enormous resources of brains, money and technology. Soon much healthier foods will be available in the United States and some other countries," the editors of *Science* magazine wrote at the close of the twentieth century.

The U.S. Department of Agriculture, which controls the direction of research and sets the standard for most conventional farmers, has been a cheerleader for genetically engineered crops since the first plant with a functioning foreign gene was created in 1983. Like the food and biotech industries, the USDA has been funneling research money into plant genetic engineering projects at the nation's major land grant universities.

In recent decades American agriculture groups, too, have wielded tremendous influence over international trade policy. In 1996 U.S. exports of agricultural products totaled $60 billion, a sum that contributed to an agricultural trade surplus of $30 billion.

By 1997, when European retailers balked at accepting U.S. shipments of gene-altered corn and soybeans, the Grocery Manufacturers of America, a trade group representing brand-name companies with sales of more than $400 billion a year, had a major stake in biotechnology. U.S. food companies use gene-altered corn and soybeans in more than two-thirds of their processed foods. Looking to the future, the food industry is hoping chemical companies will create fruits that don't spoil so quickly and flour products with extended shelf lives.

American agribusiness responded swiftly to the first stirrings of European resistance to genetically modified foods. On June 18, 1997, forty companies and influential trade groups wrote an open letter to President Clinton urging him to "get tough" with the fifteen nations of the European Union and insist that they accept America's biotech crops.

"Because trade is so important to American agriculture and the U.S. food industry, it is imperative that policy and regulations governing international commerce of genetically modified food and agricultural products are based on sound science and not just emotion which often turns into pure hyperbole," the agribusiness groups wrote. They added that "segregation of bulk commodities is not scientifically justified and is economically unrealistic."

Monsanto viewed labeling and segregation issues as crucial to the financial viability of its soybeans. American biotechnology companies, led by Monsanto and backed by the Clinton administration and the food processing industry, insisted it would be too expensive and impractical to separate gene-altered soybeans from regular soybeans. Privately, food industry officials acknowledged that labeling would be the kiss of death for biotech foods. Biotechnology companies argued that affixing a label stating "genetically engineered" to a food product would be equivalent to plastering it with a skull and crossbones symbol. Industry groups assumed that given a choice, most consumers would not buy or eat genetically engineered foods.

"It is critical that the EU understand at the highest level that the U.S. would consider any trade barrier of genetically modified agricultural products, be it discriminatory labeling or segregation, unacceptable and subject to challenge in the World Trade Organization," the letter to Clinton added.

Douglas Parr's perception that UK activists faced a formidable foe was on the mark. The power and money lined up behind Monsanto's drive to push its herbicide-resistant soybeans into European markets dwarfed the resources of Greenpeace and the other consumer groups who were trying to keep the products out of Europe. On the other side of the Atlantic, though, the giant U.S. biotech industry gravely underestimated the grit and convictions of European consumers—particularly the British.

If British activists were nervous about possible unknown health risks from genetically modified foods, ecologists were worried that biotech crops would accelerate the decline in farmland species. At a seminar sponsored by the British Crop Protection Council in Brighton on November 17, 1997, British scientists talked about the effect biotech crops might have on biological diversity in the UK. Some ecologists predicted that crops genetically engineered to resist dousing by chemical weed-killers would encourage farmers to use more of the chemicals. More weed-killers, in turn, would lead to ever greater losses of weeds, insects, birds, butterflies, moths, and small animals in the countryside.

While biodiversity has interested scientists and naturalists for centuries, international diplomatic interest in the subject dates from the United Nations environment conference in Rio de Janeiro in 1992, when 192 nations signed a Convention on Biological Diversity. The goal of the treaty is to protect the irreplaceable biological resources of the planet, which provide us with natural sources of food, medicines, and immeasurable beauty. Wild species of microbes, plants, and animals—some named, and many more as yet unidentified and unnamed—have the potential to hold cures for cancer and other diseases. Preservation of biodiversity took on critical importance in the last half of the twentieth century because the world has been losing species at an astounding rate. The biologist Edward O. Wilson has estimated that three species are forever lost every hour. Some scientists say this estimate may be too high, but it nonetheless underscores the gravity of the problem.

In the British Isles, ecologists look to farmlands—hedges, ditch sides, field corners, and especially the wild, tangled hedgerows that divide the countryside into a patch quilt of fields—as vital sources of species diversity. About 79 percent of Great Britain is farmland. Of the 60 million acres in England, Wales, and Scotland, some 46 million are farmed, with about 17 million cultivated and 29 million in pastures. British farmlands have suffered a loss of diversity in wildlife species since the 1950s, due to widespread use of herbicides and insecticides, intensive farming of single crops, and the dismantling of many hedgerows that provide a habitat for many species.

"I am discouraged by the use of genetic engineering to make herbicide-resistant crops, that lock us into continuing use of herbicides," one Scottish ecologist said at the Brighton seminar. While companies and universities had planted experimental gene-altered crops in the British Isles, none was commercially available yet. The government was taking a cautious approach to the release of genetically modified organisms into the environment on the assumption that once released, they could never be recalled. But biotech companies and university researchers were investigating a number of gene-altered crops at test plots throughout the UK, including sugar beets, corn, oilseed rape, wheat, sunflower, potato, and tomato plants. The majority of experimental plants were engineered to withstand a company's specific brand of chemical weed-killer. Monsanto led the pack in this area, altering oilseed rape plants (bred to produce the crop known as canola in North America), cotton, corn, and sugar beets to resist Roundup. A broad-spectrum weed-killer, Roundup destroys all of the wild flowers and other wild plants that spring up between rows of crops and around the edges of cultivated fields. Yet it is the wild plants, hedgerows, and brush that encourage diversity to flourish, by attracting and harboring a host of small animals, insects, butterflies, and birds.

The English are wildly enthusiastic bird-watchers. About one million people in the UK, or one in fifty, belong to a birding society. But the use of pesticides and weed-killers, coupled with reduced mixed farming, has led to a drastic drop in many songbird species. Stephen Baillie, an ornithologist with the British Trust for Ornithology in Norfolk, found that between 1970 and 1990 the number of blackbirds on farmlands declined by 40 percent. Skylarks declined by 49 percent. The song thrush saw a 65 percent drop between 1975 and 1993.

Counts of other bird species have dropped sharply as well, including the donnock, linnet, bullfinch, yellowhammer, cirl bunting, reed bunting, corn bunting, grey partridge, corncrake, mistle thrush, stone curlew, barn owl, and turtle dove.

Some ecologists at the Brighton seminar worried that through cross-pollination genes from bioengineered crops might spread gradually into related species. Over time biotech traits could be passed on in their new hosts, replicate, and migrate into yet other closely related plant species—perhaps changing the ecosystem in subtle, unpredictable ways. Biotech genes for traits such as insect resistance might confer an advantage over wild relatives of a gene-altered plant, possibly leading to the gradual elimination of natural varieties. One scientist from the John Innes Centre in Norwich said that genes inserted into crop plants definitely will move into wild species—a phenomenon scientists call "gene flow." The threat of gene flow is greatest between closely related plants. If genes conferring resistance to weed-killers migrate into weeds, environmentalists warned, the resulting "super-weeds" could be hard to destroy.

Biotechnology companies are also creating crops with "stacked genes." In this process, they insert a number of different genes into a single plant. Each gene is aimed at producing a desired trait—such as the ability to withstand a weed-killer, to kill one or more insect pests, or perhaps to resist a fungus.

The day after the seminar on biodiversity, thousands of businessmen, scientists, and government officials converged on Brighton for the start of the British Crop Protection Council's annual international meeting on agricultural chemicals. Dennis Avery, director of global food issues at the conservative Hudson Institute in Indianapolis, Indiana, gave the conference keynote address. In his talk, Avery argued that the biggest threat to world wildlife is neither pesticides nor population growth but loss of habitat, particularly from "the potential plow-down of much of the world's remaining forests to produce low-yield crops and livestock." The only way to save habitat, yet feed a growing, affluent world population with meat and eggs, he contended, is by more intensive farming of currently farmed land, using more pesticides and biotechnology.

Avery touted the theoretical power of genetic engineering to dramatically boost yields of food crops. "Biotechnology is the biggest piece of knowledge on humanity's shelves which we have not yet fully exploited. Indeed, we are just beginning to understand its power."

Rather than relying on the plants that farmers have cultivated and nurtured for centuries to feed us, Avery urged scientists to try to invent more perfect plants. "We should now be mapping the best gene groups for each trait, from all the genes, in all the species," he said. "In effect, we can construct the perfect plant from the ground up."

Avery called Bt corn and Roundup Ready soybeans "wonderful examples of high-yield technologies which use some of the safest and most sustainable technologies ever tested by science." He also praised the generation of pesticides that were invented to replace DDT, saying, "Today, modern pesticides, such as glyphosate and the sulfonylureas are applied at low rates, a few ounces per acre, are approximately as toxic as table salt and degrade from the environment in a few weeks. Moreover, these safe, effective chemicals are vastly safer for the environment than biological pest controls."

While new pesticides such as Roundup are certainly safer for humans than many of the older cancer-causing compounds, many environmental groups have argued in recent years that organic farming systems are better still, since they build healthy soils and require no synthetic chemicals. These groups believe that organic farming techniques could bring about a sustainable agriculture capable of feeding the world's population. Avery discounted the likelihood that organic farming and vegetarianism, both of which have grown in popularity in wealthy Western countries, will help to solve the looming global food production crisis. "If population growth stopped this hour, we would probably have to double the world's farm output just to provide the meat, fruit and cotton that today's 5.9 billion people will demand in 2030 when virtually all will be affluent," he said. "Humans might be able to meet their nutritional needs with less strain on farming resources by eating nuts and tofu instead of meat and milk. So far, no society has been willing to do so."

The following evening, I had a chance to hear the views of a wide range of scientists and laypersons during a public debate on the future of genetically modified foods in Europe. The lively discussion took place at the Brighton Center, a boxy, concrete building on the English Channel. The room was packed, prompting the moderator to remark, "You can fill a room on this subject any time of the day or night."

Several people complained bitterly that Monsanto and other global corporations were forcing genetically modified soya and maize on the public. By intentionally mixing genetically engineered and conventional varieties, they said, the company had prevented consumers from exercising choice. "We have to be aware that in the UK, and more so in Europe, there are whole value systems based on personal choice. Allowing consumers to act on choice is something the industry has to take on board and hasn't until now," one member of the audience said.

A middle-aged man questioned whether "the genetics we learned in university apply to transgenes. Is there any pattern of where the genes go?" he asked. "This starts to get at the guts of some of the fears of the public."

"I don't understand why we need these genetically engineered foods at all," one woman said, her tone betraying frustration and anger.

A biotechnology company official argued that the public needed to be better educated about the benefits of genetically modified foods. But an official with the Ministry of Agriculture responded bluntly that "one of the most vociferous groups against the GM technology are the middle class, educated consumers."

A middle-aged woman summarized the worries of British citizens most succinctly when she observed, "There are profits to the companies, and benefits to the farmer, but any risks will be borne by society as a whole."

Throughout the winter months of 1997, mistrust of genetically modified foods became deeply rooted among the population of the British Isles. Following the spring 1998 planting, a rebellion erupted in the fields. In Scotland, Ireland, and England, doctors, lawyers, housewives, chefs, teachers, and students joined environmental activists in digging up genetically modified crops planted by Monsanto, the French-German company Aventis, and the Swiss chemical and pharmaceutical giant Novartis. What had started a year earlier as isolated acts of vandalism against gene-altered crops blossomed into a widescale revolt.

On the night of June 3, 1998, several men and women carried scythes and other common gardening tools into a field in Derbyshire.

Working in the dark, they methodically dug up Monsanto's genetically modified oilseed rape plants and destroyed them. The nighttime scene was repeated at six other farms across England, located in Worcestershire, Gloucestershire, Lincolnshire, and Nottinghamshire.

The following morning one of the Derbyshire "croppers," as the protesters called themselves, telephoned Jacklyn Sheedy of the London-based Genetic Engineering Network and left this message: "If ever there was a clear case of agrochemical companies placing their profits before the safety of our food and our environment, genetic engineering is it. Nobody has asked for these Frankenstein foods, most people actively dislike them, and yet we are not even being consulted on what goes into our mouths. This morning, due to responsible and peaceful direct actions, we are seven fields closer to a safer world."

The anonymous activists got a boost for their cause a few days later when Prince Charles condemned agricultural biotechnology. The prince of Wales is an enthusiastic organic gardener who has long opposed researchers' use of recombinant DNA techniques to tamper with humanity's food supply. In an article in the *Daily Telegraph* of June 8, 1998, he urged scientists to stop playing God. Genetic engineering "takes mankind into realms that belong to God and God alone," he wrote. "We simply do not know the long-term consequences for human health and the wider environment of releasing plants bred this way."

A few nights after Prince Charles voiced his opposition to genetically modified foods, ten citizens stole onto a half-acre field in Kirby Bedon, a village southeast of Norwich. Under cover of darkness, they uprooted and bagged an entire crop of Monsanto's experimental genetically engineered sugar beet plants.

It was the Boston Tea Party in reverse. On Saturday, July 4, 1998, the popular revolt in England moved from darkness into the light of day. Five British women clad in protective clothing were arrested at eleven o'clock in the morning, after pulling up almost two hundred genetically engineered oilseed rape plants at Model Farm in Watlington, Oxfordshire. The women sealed the plants in bags marked with a biohazard symbol. Before the raid they sent letters to the farmer, to Monsanto, and to the police declaring their intention to destroy the genetically modified crop.

Kathryn Tulip, the first woman to be arrested, exhorted onlookers, saying, "I hope that my actions will encourage other ordinary people to join with us to take responsibility for stopping this technology from destroying our environment and endangering our food and our health."

Rowan Tilly, Zoe Elford, Melanie Jarman, and Jo Hamilton, the other members of the party, posted a press release on the Internet, announcing the launch of a new "genetiX snowball" campaign. They called for a five-year moratorium on the release of gene-altered crops in the UK. The nonviolent campaign of "civil responsibility," as the protesters described it, was to be modeled on the antinuclear movement's "Snowball campaign" of the late 1980s, in which thousands of UK citizens participated.

Tilly said, "These GE crops are an assault on our food and the environment. In the face of all responsibility being waived by those in a position to wield it, the responsibility falls on us."

Following the July Fourth genetiX field action, more than two hundred people phoned the group and offered to join in similar actions. Among the volunteers were teachers, lawyers, and a popular television chef.

Throughout the summer, British newspapers on the Internet and activists' websites carried vivid accounts of citizens' rallies and raids on plots of genetically engineered plants. With a simple click of the mouse, local activists and sympathizers from Europe to Asia were able to follow the turbulent events taking place across the English countryside. Anti–genetic engineering protesters created networks to rapidly inform one another of upcoming field actions. One environmental group posted on the Internet a complete list of the UK government's registry of permits for experimental gene-altered crops. The list included town names and specific locations for every farm where the crops were to be planted.

In Wiltshire, at six-thirty on a Friday evening in mid-July, a dozen men, women, and children wearing protective suits, masks, gloves, and footwear entered a field and pulled up genetically engineered oilseed rape belonging to Monsanto. Dozens of supporters looked on during the act of civil disobedience.

Angela Goodman, one of the participants in the action, told a local newspaper, "Contrary to recent press reports, we and others before us are neither 'eco-warriors' nor 'super heroes.' We are simply concerned mothers, fathers and other ordinary people that are unwilling to accept Monsanto force-feeding us genetically engineered crops."

Monsanto officials appeared to be flabbergasted by the angry reception their genetically engineered crops were receiving in Great Britain. The more pliant—and, it must be acknowledged, less informed—American public had seemingly accepted the genetic alteration of corn and soybeans with barely a whisper of dissent, save for protests by a handful of environmental activists. The global corporation had apparently expected a similar response from European citizens. Monsanto embarked on an expensive public relations campaign in an attempt to rehabilitate its image in the UK. In June 1998 the company launched a $1.5 million advertising blitz to convince British consumers that its herbicide-resistant soybeans and insect-killing corn were safe to eat and beneficial for the environment. The campaign backfired, however, as consumer groups accused the multinational corporation of foisting misleading propaganda on the British public.

It seemed Monsanto couldn't utter a word or print a piece of literature without provoking the ire of the British public. Even the company's use of photographs of third world children on its website and in promotional pamphlets angered English activists. Some consumer activists accused the multinational corporation of cynically exploiting images of the world's poor and hungry to promote its own profits. Monsanto was asserting that its genetically altered foods would feed the world's growing population, yet, critics charged, neither Roundup Ready soybeans nor Bt corn had demonstrated higher yields than conventional varieties of soybeans and corn. Subsequent studies at the University of Nebraska showed that Roundup Ready soybeans actually yield slightly fewer bushels per acre than conventional varieties.

Like the Irish and English, many consumers on the Continent vigorously opposed the influx of genetically engineered foods from the United States. European consumers from France to Austria also were incensed by the gene-altered soybeans and corn that began arriving at their ports unlabeled and mixed with conventional grains in 1996. In Germany environmental activists claimed responsibility for destroying at least twelve fields of experimental genetically engineered crops in 1996.

On January 27, 1997, hundreds of activists coordinated by Greenpeace demonstrated in front of multinational food companies in Austria, Belgium, the Czech Republic, Finland, France, Germany, Italy, Spain, and Switzerland, demanding they stop using genetically engineered soy in their food products. In Austria 1,226,551 citizens—about one-fifth of the nation's adults—signed a petition in April 1997 calling on the government to stop releasing GMOs into the food chain and the environment.

In January 1998 in Nerac, a group of French farmers from the Confédération Paysanne destroyed genetically engineered maize in grain silos owned by Novartis. José Bové, a passionate and articulate member of the group, explained his actions in an account circulated on the Internet. He argued that the farmers had "no other choice" but to destroy the genetically engineered corn. "The way in which genetically modified agricultural products have been imposed on European countries didn't leave us with any alternative," he wrote. "Why refuse something which is presented as 'progress'? It's not because of old-fashionedness, or regret for the 'good old days.' It's because of concern for the future, and because of a will to have a say in future development."

Citizens in all fifteen nations of the European Union pressed their governments for mandatory labeling of biotech foods. On May 26, 1998, the EU agricultural ministers adopted rules requiring labels for foods made from genetically engineered soybeans and corn. Many consumer groups were still angry, though, claiming the labeling rules did not go far enough. Under the law, the European Commission was to draw up a "negative list" of foods that would be exempt from the labels. Soybean oil and hydrolyzed starches such as glucose and fructose syrup, which are highly processed, were expected to be placed on the list of labeling exemptions.

In June 1998 I traveled to Brussels to learn more about the labeling law. One EU official told me that beyond labeling, consumers wanted a system that would allow them to trace the origins of the ingredients in their food purchases, to be certain they came from traditional, nonengineered sources.

European food company executives were carefully watching a move recently taken by the grocery chain Iceland, which had seven hundred stores in the UK. Iceland had decided to develop a complete line of foods for mass market that were guaranteed GMO-free. The action would affect about four hundred product lines, from chocolates to burgers. A Gallup poll commissioned by Iceland found that 77 percent of customers who had heard of genetically modified foods had reservations about them, and 81 percent were concerned they might already be eating foods containing gene-altered ingredients.

"We believe customers need to be informed about what is going on and need to be given a choice as to whether or not they buy GM products," Bill Wadsworth, technical manager for Iceland Frozen Foods, said at an industry meeting on genetically modified foods in Brussels in June 1998.

Iceland went to great lengths to keep gene-altered commodities out of its house-brand foods. In some cases the company had to remove the soya material from food products altogether, replacing soy oil with canola oil, for instance. Iceland found a major producer in São Paulo, Brazil, to provide nonengineered soy products, including oil, lecithin, and protein isolates. Samples of the GMO-free products would be sent to Genetic ID in Fairfield, Iowa, for testing. The Iowa testing company would then perform DNA analyses on the foods to make certain that they contained no genetically engineered proteins or DNA from genetically modified plants.

Wadsworth emphasized that Iceland's foods were not intended for an elite market such as the organic produce market. Rather, the company was intent on obtaining GMO-free soy flour for the same price as flour in which Roundup Ready soybeans and traditional soybeans were commingled. The company believed that working-class people should have the same rights to GMO-free foods as more affluent customers, and people saw no reason to pay more for traditional foods than engineered foods.

On June 25, 1998, the European Commission held a Round Table discussion on the future of biotechnology in Brussels to solicit the views of a wide swath of Europeans—from industry and academia to environmental and consumer groups. The official goal of the meeting was to foster the idea that biotechnology would play a vital role in im-

proving the future quality of life for Europeans. John Battle, UK minister for science, energy, and industry, predicted that within two years
all new pharmaceutical products would have some biotechnology
component. About 400,000 Europeans were already employed in biotechnology industries, a number that would surely increase rapidly, he
told the Round Table. Attempts to steer the discussion toward the
benefits of biotechnology in medical uses faltered, however, as consumer representatives repeatedly interjected their concerns about the
proliferation of GM foods.

Benedikt Haerlin, lead architect of Greenpeace International's
campaign against genetically modified foods and a former German
member of the European Parliament, insisted the government should
require companies to provide GMO-free soya and maize. "I'm used to
eating certain foods and want to keep doing it, and not pay extra
cost," he said, entreating the government to step in and require the
segregation of genetically modified and conventional foods.

Jim Murray, director of the Bureau of Consumer Unions, told the
Round Table that the rapid proliferation of genetically modified ingredients on the market had provoked the greatest outpouring of anger
from European consumers he had ever seen. Consumers were afraid
that GMO-free foods would disappear from the marketplace altogether and that they would have no alternative but to eat gene-altered
foods. The public was in urgent need of choices and information, neither of which they have now, he said.

"Do not force-feed consumers," Murray implored industry executives and government leaders. "Consumers are not geese!"

TWO

ALTERED STAPLES

As the groundswell of opposition to genetically modified foods rumbled across Europe in 1998, a small but earnest group of American scientists, religious leaders, and concerned citizens voiced their own discontent with biotech foods. On May 27, 1998, a coalition of rabbis, Christian clergy, biologists, and consumers sued the U.S. Food and Drug Administration for failing to require safety testing and labeling of genetically engineered foods. The lawsuit charged that the FDA's policy endangers public health and violates the religious freedom of individuals who wish to avoid foods that have been engineered with genes from animals and microorganisms.

"Genetic engineering is the most radical technology ever devised by the human brain, yet it has been subjected to far less testing than other new products," Steven Druker, president of the Alliance for Bio-Integrity, the lead plaintiff, said at a press conference in Washington, D.C., to announce the lawsuit. "This lax regulatory process is scientifically unsound and morally wrong."

Druker said that Genetic ID, the DNA testing company in Fairfield, Iowa, had tested several soy-based infant formulas in 1997, looking for genetically engineered ingredients. Four baby formulas containing soy ingredients had tested positive.

FDA officials had always maintained that genetically engineered foods were substantially the same as their conventional counterparts. Yet, the plaintiffs argued, if the foods contained new constituents that could be detected and measured, how could government regulators assert they were virtually identical to traditional ones?

"Because of FDA's failure to require labeling, millions of American infants, children and adults are consuming genetically engineered food products each day without their knowledge," charged Andrew Kimbrell, director of the Center for Food Safety in Washington, D.C., another plaintiff in the lawsuit. The center is a small, national membership group set up in the early 1990s to assess the human health and environmental safety of new food technologies. "The FDA has made consumers unknowing guinea pigs for potentially harmful, unregulated food substances," he told the large group of reporters assembled for the morning press conference at the National Press Club, a few blocks from the White House.

A central issue in the lawsuit involved consumers' rights to know about the new genetic material being engineered into their food. Druker pointed out that because the FDA requires neither labeling nor pre-market approval of gene-altered foods, American consumers have no way of knowing what genetically modified foods are on the market. To the best of anyone's knowledge, he offered, by mid-1998 the FDA had allowed at least thirty-six gene-altered foods to enter interstate commerce without labels and without proper analysis of the potential safety risks associated with their genetic instability.

Most of the thirty-six foods named in the lawsuit were crops that had been genetically engineered either to kill insect pests or to resist a particular company's brand of herbicides. Foods in this category included genetically modified corn, canola, cotton, potato, and soybean varieties from AgrEvo, Monsanto, and Novartis. Three other kinds of genetically engineered foods were listed in the lawsuit as well: virus-resistant papayas developed by researchers at the University of Hawaii and Cornell University, virus-resistant squash from Seminis Vegetable Seeds and Asgrow Seed Company, and tomatoes modified for changes in ripening from Monsanto, Agritope, and Zeneca.

Rick Moonen, a New York City chef and owner of two restaurants, Oceana and Molyvos, explained that his support for the lawsuit grew out of the obligation he feels to his patrons to know precisely what is in the foods he serves them. "People come to Oceana because they trust me. They know that I'm going to source out the highest-quality ingredients in the market for their dining experience," he said. "By not requiring mandatory labeling and safety testing of all genetically engineered foods, the government is taking away my ability to assure customers of the purity of the foods they eat at my restaurant."

A packet of press materials that the litigants had assembled included a simulated dinner menu of genetically engineered foods. Some of the transgenic plants and animals on the mock menu were already on the market. Others were products of federally funded research or crops that have received permits from the USDA for environmental testing:

APPETIZERS

Fingerling Potatoes with Waxmoth Genes
(served with sour cream from bovine growth hormone treated cows)
Juice of Tomatoes with Flounder Genes

ENTREE

Braised Pork Loin with Human Growth Genes
Boiled New Potatoes with Chicken Genes
Fried Squash with Watermelon and Zucchini Virus Genes
Toasted Cornbread with Firefly Genes

DESSERT

Rice Pudding with Pea and Bacteria Genes

BON APPETIT!

The menu illustrates plainly enough why the food industry has adamantly opposed consumer efforts to require labeling on genetically altered products. Given the choice between traditional potatoes and potatoes spiked with waxmoth genes, how many people would select the potato-insect variety?

Religious leaders joining the lawsuit insisted that the FDA's failure to regulate and label gene-altered foods violates the free exercise of religion by Americans who wish to avoid such foods for religious or moral reasons.

"Everyone who believes the biosphere developed through the purposeful plan of a benevolent God should reject gene-altered food to preserve the dignity of that plan," said Rabbi Harold White, a lecturer in theology at Georgetown University. "Since the dawn of life on Earth, divine intelligence has systematically prevented transfers between widely differing species. Limited human intelligence should not rush to make such unprecedented transfers commonplace."

Steven Druker believes that gene-altered foods violate Halakha, Jewish law—a belief not universally shared by Jewish thinkers. Born and raised in Des Moines, Druker left Iowa in 1964 to attend the University of California at Berkeley. He received a law degree from Berkeley's School of Law in 1972, then practiced law in San Francisco and Los Angeles for a number of years before returning to Iowa. It was in the mid-1990s, while researching a book on ethics and spirituality, that Druker learned about what he calls "the genetic reconfiguration of the food supply."

In 1997 he wrote a paper entitled "Are Genetically Engineered Foods in Accord with Jewish Law?" that examines the finer points of Jewish dietary law. For many observant Jews, the transfer of pig genes into fish or vegetables, for instance, would raise obvious concerns about the resulting genetically altered foods. "Organisms implanted with genes from non-kosher species are themselves non-kosher and must be avoided," Druker concluded. In his interpretation of Jewish law, any rearrangement of genes between species is worrisome. For purposes of food production, he argued that "transposing genes between species that are naturally prevented from cross-breeding is a high-risk venture" that is unsound under Jewish law even when both species are kosher.

FDA officials actually did give some consideration to the ethical and religious implications of inserting animal genes into plants, as I discovered in searching through agency documents. James Maryanski, biotechnology coordinator of the FDA's Center for Food Safety and Applied Nutrition, had a hand in developing the agency's policy on biotech foods in 1992. Maryanski gave the following explanation of why the FDA believes that moving genes from one organism to another should not trouble people for religious, ethical, or moral reasons.

"There are thousands of genes in a plant. When a scientist adds new genes from an animal, it gives that plant several new proteins. But these proteins would not seem to give animal characteristics to the vegetable." One point people are not generally aware of, Maryanski continued, is that "there are genes in humans and animals that are in plants. There is a gene tha Pt occurs in rice that also occurs in the human brain. Vegetarians would not avoid rice because of that. Our current view is that these modifications will not result in foods that violate any ethical or religious considerations."

To many, Maryanski's explanation is an unsatisfactory response. The FDA's policy simply declares that genetically engineered foods are substantially equivalent to their traditional counterparts. It does not require that they be labeled unless a gene from a food known to cause allergies has been inserted into the new food plant. But without labels identifying genetically modified foods and ingredients, Americans who wish to avoid them for religious, ethical, or other dietary reasons are denied the right to do so.

The FDA's policy does more than violate religious laws, according to the scientists joining the lawsuit; it also abrogates good scientific principles. Philip Regal, a biologist and professor of ecology, behavior, and evolution at the University of Minnesota at St. Paul, believes "there is no sound scientific basis" for the presumption that genetically engineered foods are equivalent to, and thus as safe as, their natural counterparts. Such a presumption, he contends, "can only be made by systematically ignoring a large body of solid and relevant evidence."

There is no "general recognition of the safety of genetically engineered foods among those members of the scientific community qualified to make such a judgment," Regal wrote in testimony to the court. He recalled discussions among federal regulators at a policy meeting on biotechnology in Annapolis, Maryland, in 1988, in which "government scientist after scientist acknowledged there was no way to assure the safety of genetically engineered foods. Several expressed the idea that, in order to take this important step of progress, society was going to have to bear an unavoidable measure of risk."

John Fagan, who received his Ph.D. in biochemistry and molecular biology from Cornell University and spent seven years doing research funded by the National Institutes of Health, was another party to the lawsuit. By the mid-1990s, he wrote, in testimony submitted to the court, he had grown concerned that some applications of genetic engineering were outpacing "our knowledge about the intricacies of DNA and its interactions with the rest of the living system" and "our ability to predict and control the outcomes of genetic manipulations." In 1994, taking an ethical stand against what he considered "irresponsible trends in biotechnology," Fagan returned a $613,882 grant to the NIH and withdrew applications worth an additional $1.25 million.

"FDA's claim that we can reliably assure ourselves about the safety of a genetically restructured organism merely by knowing features of the species involved is scientifically unsound," Fagan observed. "The only way to even begin to make a reliable safety assessment of a genetically engineered organism is to rigorously test for unintended and unexpected changes. This at a minimum entails well-designed feeding studies using the whole food, not just an extract of the substance(s) known to be directly produced by the foreign genetic material."

The FDA's authority to oversee the safety of the American food supply is granted by the federal Food, Drug and Cosmetic Act first enacted in 1938. Under that law companies are required to gain pre-market approval from the FDA for food additives. Druker and the other plaintiffs in *Alliance for Bio-Integrity, et al. v. Donna Shalala, et al.,* alleged that the FDA's 1992 policy violates the act because genetically engineered foods involve "material" changes that require labeling.

The FDA defines a food additive as any substance "the intended use of which results or may reasonably be expected to result, directly or indirectly, in its becoming a component of food . . . and which is not generally recognized as safe" for such use. Food additives include such substances as flavorings, thickeners, artificial sweeteners, and preservatives. Many of the substances added to foods, including spices and food-processing enzymes that have been in the food supply for a long time, are exempted from review by the FDA because they are considered to be "generally recognized as safe," or GRAS.

If a company wants to use a new additive in a food, it must submit a petition to the FDA and provide scientific information showing reasonable certainty that the additive will be safe for the intended food use. These petitions must be publicly disclosed. The FDA has not viewed the genetic materials transferred into bioengineered foods as food additives. Instead, the agency has granted them a blanket status as GRAS.

The Food, Drug and Cosmetic Act requires labeling to differentiate all kinds of changes in foods, many of which do not affect the safety of the product. Orange juice from concentrate, for instance, must be labeled as such, to distinguish it from fresh-squeezed orange juice, even though both forms of juice are equally safe. Yet the FDA decided, categorically, that Bt corn that had been genetically engineered to make a protein that kills insects did not need to be labeled to differentiate it from regular corn.

The Grocery Manufacturers of America (GMA) issued a press release refuting the Alliance for Bio-Integrity's charges and making it clear that the food industry supported FDA's 1992 policy on genetically engineered foods. "There is no scientifically valued distinction between the safety of genetically enhanced food and food grown by traditional methods. It would be enormously impractical to label every genetically modified new crop and would falsely imply a difference in the foods' safety," said Stephen Ziller, GMA vice president. The Washington, D.C.–based group represents about 90 percent of the name-brand foods and beverages sold in the United States, with sales of more than $430 billion in 1998. Labeling genetically modified foods would imply that they are less safe or less wholesome than their traditional counterparts, the GMA insisted.

"We should not allow the opponents of progress and science-based research to turn back the clock on the extraordinary advancements we have made to feed the world and reduce farmers' dependency on pesticides and fertilizers," Ziller added.

Two years after the press conference, I talked with Joseph Mendelson, an attorney representing the Center for Food Safety, about the status of the lawsuit. In the course of our conversation, he mentioned that the lawsuit had received very little attention in the major media. One science journal that reported on the case, *Nature Biotechnology*, highlighted a comment by the Biotechnology Industry Organization calling the lawsuit "an entertaining example of guerilla theater." Most major newspapers did not report on the case until long after it was filed.

The Wall Street Journal ran an article on the lawsuit on August 18, 1999, fifteen months after it was filed. The story had the flavor of a style piece about an eccentric group of fanatics, rather than a serious news report about food safety and labeling issues. Druker's search for co-plaintiffs was described thus: "And so he began crisscrossing the country, gathering his Noah's Ark of plaintiffs, many of whom share his mystical spirituality and distrust of authority." The article, entitled "Motley Group Pushes for FDA Labels on Biofoods to Help Religious People," noted that the list of supporters consisted of 113 Christians, 37 Jews, 12 Buddhists, and 122 people who checked a box saying "my faith is not easily categorized." It made no mention of the scientists joining the law suit.

Druker found the tone of the *Wall Street Journal* article upsetting. In his view, the article showed that the newspaper had "an agenda to make the critics look bad and make biotechnology look good." Of the scant coverage major newspapers gave to the GM food labeling issues raised in his lawsuit in 1998, Druker added: "I have to ask, is there a free press?"

Like Druker, many critics of genetically engineered foods have questioned how Americans could engage in an informed debate on the risks and benefits of genetically modified foods when so few people knew they were eating them. Consumer activists often complained that major U.S. newspapers ran relatively few stories on the proliferation of biotech crops from their introduction in 1996 until mid-1999. National polls conducted throughout this time period showed that fewer than 10 percent of Americans were aware that they were eating genetically engineered corn and soybeans—figures that seemed to bear out the activists' perception. Ronnie Cummins, national director of the Organic Consumers Association in Little Marais, Minnesota, called the phenomenon "the U.S. media blackout on genetically engineered foods." A longtime anti-biotechnology activist, Cummins directed the Washington, D.C.-based Pure Food Campaign from 1994 to 1998. He set up one of the earliest domestic websites to make articles on agricultural biotechnology from various U.S. and overseas news organizations available to a wider audience. The grassroots Organic Consumers Organization, which he founded in 1998, has two hundred thousand members.

Some media analysts have suggested that the consolidation of major media outlets and corporate buyouts of news organizations in the 1990s have resulted in increased coverage of sex scandals and entertainment and decreased reporting on public policy issues. Throughout 1998, when gene-altered food crops were proliferating across America, the media displayed a frenzied obsession with Monica Lewinsky's Oval Office tryst with President Clinton. The Lewinsky scandal, which broke in January 1998 and continued through President Clinton's impeachment trial in February 1999, siphoned off the energies of untold numbers of journalists and soaked up countless inches of newspaper ink.

New York University law professor Burt Neuborne attributes the failure of the media to cover important public policy issues in part to the fact that news organizations are now owned by large conglomerates. "You don't have people controlling the press anymore with a fervent sense of responsibility to the First Amendment. Concentrating on who's sleeping with whom, on sensationalism, is concentrating on essentially irrelevant issues," he said during a *Columbia Journalism Review* forum.

Other analysts suggested that small news organizations were reluctant to report on food safety questions for fear of invoking the wrath of farm trade groups or biotech companies. Thirteen states have passed so-called "food disparagement laws" aimed at protecting farm groups and food companies from media reports that might cause consumers to avoid their products. "Though some publishers and broadcasters continue to put out new reports on food, other media companies, especially smaller ones worried about the high cost of defending a lawsuit, have stricken information from manuscripts, avoided certain food issues, or, in one case, dropped a book project that was already at the printer," Melody Petersen reported in a *New York Times* article on the chilling effect that food disparagement laws have had on freedom of speech.

Texas cattlemen sued Oprah Winfrey for a remark she made on her television show in April 1996. Her guest had talked about the threat that mad cow disease might potentially pose to America's beef industry. (The disease has never been found in cattle in the United States.) Winfrey said the information stopped her "from eating another burger." In February 1998 a jury in Amarillo, Texas, ruled in Winfrey's favor, but she spent more than $1 million on legal fees, and the cattlemen appealed the decision. In February 2000 the Fifth Circuit Court of Appeals upheld the lower court verdict.

The government's advocacy role in promoting genetically engi-

neered foods might also have played a role in the sparse coverage of
biotech corn and soybeans before 1999. Officials at the FDA and
USDA had consistently treated biotech foods as essentially the same as
regular foods, suggesting to inquiring reporters that 'there's no story
here.' In March 1997 Agriculture Secretary Dan Glickman stated,
"Biotechnology's been around almost since the beginning of time. It's
cavemen saving seeds of a high-yielding plant."

When genetically engineered corn and soybeans were introduced in
1996, government officials made no efforts to notify the public about
them. The FDA, USDA, and EPA are funded by billions of dollars of
taxpayer money and could easily have launched a major public out-
reach campaign to tell consumers about the new foods. Whatever the
cause—editorial neglect, corporate influence, government silence, or a
combination of factors—Americans were kept in the dark about genetically engineered foods.

In April 1998, though, a high-profile lawsuit brought the issues
of corporate influence over the news and genetically engineered foods
into sharp focus. Two award-winning investigative reporters in Tampa,
Florida, sued Fox News, alleging that the station pressured them to report false and misleading information to the public about a genetically
engineered product—Monsanto's bovine growth hormone.

In 1993 the FDA approved a controversial animal drug made by
Monsanto called recombinant bovine somatotropin, or rBST. The
drug is also commonly called recombinant bovine growth hormone,
or rBGH. The drug significantly boosts milk production in cows. The
extra milk sometimes comes at a cost to the health of the animals,
however. Critics have charged that cows sometimes develop infections
such as mastitis, which in turn can cause their milk to contain unacceptable levels of pus, unless the cows are treated with antibiotics.

When rBGH was first brought to market, a spate of demonstrations
erupted across the United States. The country already had an oversup-
ply of milk, critics argued, so what was the point of introducing a potentially hazardous drug to boost milk production? Further, the FDA
had decided that the hormone-treated milk would not be labeled,
which meant that consumers would not be able to distinguish it from

regular milk. In 1994 angry protesters poured milk into the streets. Gradually, though, the protests died down, as did news coverage of the genetically engineered hormone, which Monsanto sold under the brand name Posilac. American dairy farmers continued to administer the drug to their herds, however—and a small percentage of consumers began buying organic milk and searching for ice cream and other dairy products that were advertised as being made from milk produced without rBGH.

The issue gained public attention again in December 1997, when Fox Network–owned WTVT in Tampa, Florida, fired Steve Wilson and Jane Akre, two highly acclaimed investigative reporters. The husband-and-wife team had produced a series of news reports on the widespread use of Posilac in Florida's dairy herds. The four-part series had been set to air on February 24, 1997, and would include the views of a university researcher who warned that people faced possible health risks in drinking milk from cows treated with rBGH. Fox was so excited by the series that the station bought considerable radio time to advertise it. The series was billed as "The Mystery in Your Milk." But on the Friday before the scheduled Monday broadcast of the first report, Fox News chairman Roger Ailes received a letter from a Monsanto lawyer warning the network not to air the story.

"There is a lot at stake in what is going on in Florida, not only for Monsanto but also for Fox News and its owner, as well as for the American people and a world population that can benefit significantly from the use of rBST and other products of agricultural biotechnology," John J. Walsh, a New York attorney with the firm of Cadwalader, Wickersham & Taft, wrote. The letter also charged that the two journalists had "no scientific competence." The series was delayed for a week, then was postponed again after Fox received another letter from Monsanto's attorney. The second letter, dated February 28, 1997, warned that some of the points to which reporter Jane Akre was seeking a response from Monsanto contained "the elements of defamatory statements which, if repeated in a broadcast, could lead to serious damage to Monsanto and dire consequences for Fox News." Over the next seven months, the journalists rewrote the series no fewer than eighty-three times, but none of the rewrites met the station's approval.

Before he joined the Tampa station, Steve Wilson had spent five years as a senior investigative correspondent for the TV news-magazine *Inside Edition*. He came to Fox with four Emmy awards for his investigative work and twenty-six years experience at CBS, ABC, and elsewhere. Akre has worked as a television news reporter and anchor for more than twenty years, including working for CNN in the mid-1980s. When WTVT hired them in November 1996 to produce hard-hitting investigative reports, they were billed as a "dream team" in the local Tampa market. As Wilson later put it, "Our dream turned into a journalist's nightmare."

Wilson said that Fox threatened to fire him and his wife if they would not make the story conform to the network lawyers' wishes. He told the television station that the couple would file a formal complaint with the Federal Communications Commission for deliberate news distortion. In response, Wilson said, "We were offered nearly $200,000 to quietly go away and say nothing about the TV station's handling of the story." Akre and Wilson said they refused the cash and were let go when their contract came up for renewal in December 1997.

The television series Akre and Wilson produced contained information suggesting a possible link between cancer and milk from cows that had been injected with Monsanto's rBGH. Some studies have suggested that cows treated with rBGH produce higher levels of insulin-like growth factor 1, or IGF-1, than do untreated cows. The growth factor is a normal hormone protein that occurs in both people and cows. Walsh stated in his first letter to Fox chairman and CEO Ailes, "The opportunities for misinterpretation of scientific inquiries into this question are enormous, particularly when engaged in by lay persons such as Akre and Wilson. The key point you should understand is: there is no increase in the level of IGF-1 in the milk which comes from rBST treated cows."

Samuel Epstein, a researcher with the University of Illinois in Chicago, studied the relationship between higher levels of insulin-like growth factor 1 in people and the incidence of cancer. He concluded in 1996 that IGF-1 from hormone-treated cows may increase people's chances of contracting breast cancer and colon cancer. "With the complicity of the FDA, the entire nation is currently being subjected to an experiment involving large-scale alteration of an age-old dietary staple by a poorly characterized and unlabelled biotechnology product," Dr. Epstein wrote in a paper on his findings. His warnings were featured in Akre and Wilson's series on Posilac-treated dairy cows in Florida.

Shortly after Fox fired Akre and Wilson, a report published in a respected science journal bolstered Epstein's views. In January 1998 *Science* magazine reported that men with high blood levels of IGF-1 are four times more likely to develop prostate cancer than are men with low levels of the hormone. The article said that IGF-1 stimulates the division and proliferation of both normal and cancerous prostate cells.

Then in May 1998 another prestigious scientific publication, the British journal *Lancet,* reported that premenopausal women with higher blood levels of IGF-1 are seven times more likely to develop breast cancer than are women with lower levels of IGF-1.

"Scientists have found that the IGF-1 levels in the hormone-treated milk could be 70 percent more than in regular milk. That's a risk I am personally not willing to take," Jane Akre told me later. "Why should I take a risk with the milk I feed my child? What's the upside for me, as a consumer, to take any risk?"

Monsanto maintains that milk from cows treated with Posilac is the same as normal milk. "There is no difference," Monsanto spokesman Gary Barton told me in an interview for *Food Chemical News* in December 1998. He added that there is no test that can distinguish between milk from rBGH-treated cows and regular milk.

Monsanto's claim is fully backed by the FDA, whose officials say the hormone-treated milk is perfectly safe and virtually identical to normal milk. But Posilac has been banned in the fifteen nations of the European Union. Canadian regulators have refused to approve use of the drug because they are not convinced it is safe.

In 1998 a team of scientists in the Health Protection Branch of Health Canada conducted a review of the data on which the FDA had relied to conclude that the synthetic growth hormone is safe. The Canadian scientists raised many questions about "gaps" in the ninety-day rat study that Monsanto submitted to the FDA in 1990 as evidence of the drug's safety. "The 1990 evaluation was largely a theoretical review taking the manufacturer's conclusions at face value. No details of

the studies nor critical analysis of the quality of the data were provided," the Canadian scientists wrote in a report on their review.

The Canadian team suggested that a three-month study of the hormone's effect on a nonrodent species such as a dog should have been
required, as well as definitive studies to demonstrate that rBST and
IGF-1 are not absorbed after oral intake.

At the end of 1998, in response to the troubling findings of the
Canadian team, a coalition of more than twenty U.S. citizens' groups
and environmental organizations petitioned the FDA to remove rBGH
from the market until additional safety tests were performed on the
drug. The petition argued that when the FDA approved rBGH as safe,
the agency had asserted that "no toxicologically significant changes
were noted" in rats administered rBGH orally.

At a December 15, 1998, press conference called by the petitioners,
Michael Hansen, a researcher with Consumers Union in Yonkers,
New York, pointed out that the Canadian reviewers found that between 20 and 30 percent of the rats in Monsanto's ninety-day study
developed "an antibody response" to the drug. Some of the rats also
developed cysts. "These are toxicologically significant changes in the
rats, and they should have triggered a full human health review, including assessment of the potential carcinogenic and immunological
effects," Dr. Hansen told reporters.

"It is clear that the FDA put the interest of Monsanto above its duty
to protect the health of the American consumer when it approved
BGH five years ago," charged Andrew Kimbrell, who directs the
Washington, D.C.–based Center for Food Safety, summarizing the position of the twenty consumer groups.

Vermont's two senators also felt that the Canadian report raised
significant questions about the FDA's finding that rBGH is safe for
humans. In December 1998 James Jeffords, then a Republican, and
Democrat Patrick Leahy wrote to Secretary of Health and Human
Services Donna Shalala asking her to consider whether any new regulatory action on FDA's part is needed.

On December 14, 1998, Monsanto sent a letter to a staff member
in Jeffords's office, attesting to both the safety and the popularity of
rBST among dairy producers across America. "Posilac is the largest
selling dairy animal health product in the U.S.," the biotech company

wrote. "Approximately 13,000 producers, representing 30% of U.S. dairy cows, are using Posilac." The company added that Posilac's market penetration in Vermont was consistent with the national average of 30 percent. Monsanto noted that in 1998 the drug was also sold abroad, to dairy producers in Egypt, Israel, Jordan, Korea, South Africa, Turkey, and several Latin American countries.

The FDA declined to change its position on the safety of rBGH.

In August 2000, two years after Akre and Wilson sued Fox News, charging the station with pressuring them to broadcast false and misleading information about Monsanto's bovine growth hormone, a jury in Tampa ruled in their favor. At the end of a five-week trial, the jury found that Fox had, in fact, taken retaliatory action against Akre when she threatened to blow the whistle to the FCC. She was awarded $425,000 in damages. The jury did not find for Wilson, apparently concluding that Fox's decision not to renew his contract was not based solely on a threat to blow the whistle to the FCC. Wilson said the couple shared in the victory nonetheless.

Fox, Wilson, and Akre have all appealed the decision, which only accorded Akre a partial victory. Lawyers with the firm of Williams & Connolly, representing Fox, argue that Akre failed to establish that the station acted deliberately to falsify the news "about any matter affecting the basic accuracy of the news report." The Fox appeal also raises a technical problem with the timing of her written notice of a threat to disclose a violation of the FCC's news distortion policy.

Lawyers for Akre, meanwhile, argue that even though WTVT had promoted the story's investigation of a possible link between BGH and cancer, "scientific evidence suggesting such a link became off limits after Monsanto complained. Editors even avoided use of the word 'cancer,' and substituted the euphemism 'human health implications.'"

Lawrence Grossman, a former president of NBC News and PBS, fears the lawsuit will only further endanger serious investigative reporting by television stations. In the face of threats by Monsanto, WTVT's news director Phil Metlin "together with an army of news editors, station executives, and lawyers, worked with the reporters to try

to produce what one lawyer called a fair, accurate, balanced, and veri-
fiable story that would protect the station from 'risk or harm caused by
inaccuracy, carelessness, lack of balance, or perceived bias,'" Grossman wrote in the *Columbia Journalism Review.* "The shots that Monsanto had fired across the station's bow obviously struck home."

The U.S. government's handling of milk treated with Monsanto's
genetically engineered bovine growth hormone was strikingly similar
to its handling of the company's genetically modified soybeans and
corn. In both cases, the FDA declared that the new gene-altered products were "substantially equivalent" or "virtually" identical to their
conventional counterparts. In both cases, the new products were
brought to market unlabeled and were mixed in with traditional foods
and milk. And both the milk and the gene-altered foods were approved relatively quickly, on the strength of only a limited number of
animal feeding studies.

European consumers have been unwilling to take a chance on consuming either genetically engineered foods or rBGH milk. Yet these
gene-altered foods are basic staples of the American diet, consumed
daily by nearly every man, woman, and child in America. In the end,
the biotech companies and U.S. regulators may turn out to be right
about the long-term safety of these gene-altered staples. But the FDA's
review methods are suspect: the agency's habit of approving biotech
food products on the basis of a handful of industry-sponsored animal
tests appears to leave American citizens unprotected. What if the Europeans are right, and genetically modified foods turn out to cause
latent health problems?

THREE

PUSZTAI'S POTATOES

On August 10, 1998, a distinguished UK scientist gave a television interview about a study that called into question the safety of genetically modified foods. Arpad Pusztai reported finding that genetically modified potatoes cause immune system damage to rats. "We are assured that this is absolutely safe and that no conceivable harm could come to us from eating it. But if you gave me the choice now, I wouldn't eat it," the sixty-eight-year-old biochemist stated.

Pusztai had no reason to expect any problems when he began to study the health effects of genetically modified potatoes on rats. The line of gene-altered potatoes that he used had remained stable since it was created in 1992. But when he fed the potatoes to rats, he found that some of the animals had stunted growth. Their immune systems were damaged, and their hearts and livers decreased in size. Even more troubling, the brains of some of the test animals were smaller than normal.

In his 150-second interview on the *World in Action* program, Pusztai said that genetically modified foods had not undergone enough testing on animals before being approved for humans to eat. He added that it was "very unfair to use our fellow citizens as guinea pigs."

The potatoes in Pusztai's experiment were genetically engineered to produce a protein called a lectin. The lectin gene was taken from the bulb of a snowdrop plant, *Galanthus nivalis*. Snowdrops are lovely little white flowers that appear in clusters at the first sign of spring, soon after the snow melts. The lectin protein acts as a natural pesticide in the snowdrop plant, protecting it from aphids and other insects. Aphids also feed on potato plant foliage and can carry viruses that infect the tubers. Scientists in the UK and elsewhere had decided to engineer the lectin gene into potato plants in hopes it would function as a pesticide, killing aphids and nematodes that attempt to feed on the plants. Before the rat-feeding study began, UK researchers had successfully inserted the lectin gene into two lines of potato plants and then grown the genetically modified potatoes during field tests conducted between 1992 and 1995.

In his study, Pusztai fed one group of rats ordinary potatoes, a second group genetically modified potatoes, and a third group ordinary potatoes into which he mixed purified lectin proteins. The goal of the study was to devise safety testing methods to find out whether the genetically modified potatoes had any physiological effect on the mammalian gut or the nutritional status of animals. The project was funded by a £1.6 million grant awarded in November 1995 by the Scottish Office of Agriculture, Environment and Fisheries, which oversees agriculture and nutrition research programs in Scotland. The government commissioned the study because the world had paid so little attention to the effects of genetically modified foods on human nutrition. Pusztai, his colleagues at the Rowett Research Institute, and scientists from the Scottish Crop Research Institute and the Durham University Department of Biology were chosen for the grant from a field of twenty-eight bidders from across the European Union. Pusztai was elected by his colleagues to coordinate the research team.

"Everything appeared to be going along well with the potatoes," Pusztai later told me. "They resisted aphids and seemed to have a strong effect on nematodes." Before receiving the grant money, he had extracted and purified proteins from the snowdrop bulb lectin and fed the substance to rats at a concentration a thousand-fold higher than levels found in the modified potato. The lectin had only "very mild effects on the rat gut and it was counterbalanced by some beneficial effects," he said. So the scientist was taken completely by surprise when problems showed up in the genetically engineered potatoes grown for the rat-feeding study.

"I was quite certain it was impossible—it was inconceivable, as I've said—that this could have any unexpected effect. That was the reason we persuaded the Scottish Office to give us this £1.6 million. It's quite a bit of money. And our research proposal had been peer-reviewed by everybody. They don't give you that much money without having a good look at it," he added, chuckling. Arpad Pusztai spoke methodically, his English overlaid with a Hungarian accent. He is a world authority on lectins; his professional credentials are impeccable. A political refugee from Hungary in 1956, he received a Ford Foundation scholarship to study for his doctorate in biochemistry at the University of London. He did postdoctoral research at the Lister Institute of Preventive Medicine and then took a research position at the Rowett Research Institute in Aberdeen, Scotland, where he remained for thirty-five years.

Pusztai's rat-feeding study marked the beginning of what was intended to be a close examination of all aspects of the safety of genetically modified crops: their effects on soil microorganisms, the environment, beneficial insects, and animal and human nutrition. "At the time we started in 1995, there was not a single scientific publication on the potential health effects of any genetically engineered crops," he said.

In 1995 there were no commercial crops in the food chain genetically engineered to kill insects. In fact, the only genetically modified crop on the market was the Flavr Savr tomato, which was produced by genetic engineers at Calgene, a small biotechnology company in Davis, California. The Flavr Savr tomato was the first genetically modified fruit to be approved by the U.S. Food and Drug Administration. The Flavr Savr contained a copy of an "antisense" gene, which is a so-called "backward gene," for the fruit-softening enzyme polygalacturonase. The antisense gene was supposed to reduce the amount of ripening enzyme produced in the Flavr Savr tomato. This, in turn, was supposed to give the tomato an extended shelf life, since it would not soften as quickly as a regular tomato. While the law did not require a safety assessment of the tomato, Calgene officials asked the FDA to review its safety anyway. After more than three years of studying Calgene's tests and information about the gene-altered fruit, the FDA in 1994 approved the Flavr Savr as safe for people to eat.

Calgene openly introduced the Flavr Savr into the market as a new, genetically engineered food. The tomatoes were shipped to grocers along with brochures describing how they had been made. For a number of reasons, the Flavr Savr fizzled quickly in the marketplace. The tomato tended to bruise easily, which made it hard to pick, pack, and ship. Another problem, according to Robert Goodman, a former Calgene official, was that the tomato had been developed in the West, where the soil and climate were very different from those found in Florida, where the company contracted with farmers to grow it commercially. Within a year of its introduction, the first gene-altered whole food could scarcely be found in U.S. produce aisles.

To redress the lack of peer-reviewed studies on the safety of biotech foods, Pusztai's research team planned to start safety-testing genetically modified potatoes on laboratory animals. Assuming no health problems showed up in rats during the first three years of the study, they planned eventually to proceed to controlled human feeding tests.

"I think we had a quite nice, auspicious start," he told me, "because we had a product, it had been pretested, so we were very, very confident that this would be okay." As a thirty-five-year veteran of nutrition studies, Pusztai knew that the chemical composition of potatoes can vary widely, depending on the plants' growth conditions. Even small variations in the amount of sunshine and water or the position in the plot can alter the composition of a plant's fleshy tubers. To make certain that environmental factors would not taint the reliability of their experiment, the researchers took pains to grow two genetically modified potato lines and the nonmodified parent line in identical conditions, side by side, in confinement under a tent.

When Pusztai measured the nutrients in potatoes from the two genetically modified plant lines—the two "transformed" lines—he found they were not equivalent to potatoes from the parent line. Potatoes dug from one of the genetically transformed lines had 20 percent less protein than the nonmodified parent line. The modified potatoes also had "antinutritive" agents, including higher-than-expected levels of compounds that inhibit digestive enzymes. These inhibitors would interfere with nutritional processes in the stomach of rats—or humans—eating the potatoes.

Pusztai was troubled by the fact that the two genetically transformed potato plants had widely different compositions. "There were significant compositional differences between the two transformed lines, and yet these two transformed lines came from the same transformation set. That straightaway drew our attention to the fact that genetic modification may introduce unexpected changes and is not really a proper technology, because you're not reproducing exactly what you're doing," he explained. "In effect, if you use the same construct, the same vector, the same gene, everything the same, and you do this transformation again and again, you end up with different products."

The phenomenon Pusztai was describing suggested that potentially harmful changes could appear unexpectedly in the composition of genetically engineered plants several generations after the transformed plant was approved as safe for farmers to grow. If this were true, then foods harvested from genetically modified crops might always be suspect.

Pusztai knew that rats fed a diet of potatoes containing 20 percent less protein would not grow as well as rats fed more protein. In controlled animal studies, the diets of animals receiving a test substance must contain the same amount of protein and calories as those not receiving the test substance, to be considered equivalent. So he added a protein supplement to the diets of the rats fed the modified potatoes. Even after compensating for the protein, he found that the rats fed engineered potatoes had immune system damage and changes in the size of several organs.

With such unexpected health problems showing up in young animals, Pusztai said he never would recommend that the genetically modified potatoes go into a clinical trial with humans. Indeed, a finding of health problems in a rat study raises immediate concerns for researchers developing any kind of food additive or drug. A finding of immune system problems in rats might be grounds to jettison a test substance. At the very least, such a finding would trigger more stringent tests on laboratory animals.

"Let's assume that genetically modified soya is okay. I don't know, I haven't tested it, but let's just assume it doesn't show any of the things we found with GM potatoes," Pusztai continued. To test the safety of Roundup Ready soybeans on humans, a researcher would make up one diet of which a substantial part is genetically engineered soybeans and another diet with an equal amount of regular, nonengineered soybeans, he said. Each diet would be fed under controlled conditions to different groups of people, and then they would be examined for any differences.

"But none of this has been done, correct?" I asked, interrupting him.

"It has never been done," he answered. "That's the reason why I am saying we are the guinea pigs."

Pusztai's TV interview sparked an outcry from British lawmakers and consumer activists. Some members of Parliament called for a ban on genetically modified foods. Norman Baker, the Liberal Democrat Party environment spokesman, disclosed that genetically modified foods had been taken off the menu at the House of Commons. "The new health findings are very worrying and show that we have become the guinea pigs in a gigantic experiment," Baker told the BBC. "The government has been irresponsible and spineless in allowing GM foods into our diet without demanding to see definitive proof that they are safe."

Two days after the TV show aired, a controversy flared up over the scientific validity of Pusztai's study. On August 12, 1998, the Rowett Research Institute issued a statement rebutting his claims. Pusztai had "muddled" his findings, the institute charged, suggesting that the experienced biochemist had confused data from two rat-feeding experiments. Since the institute declined to release the study results, and information provided by various sources was vague and fragmented, it was hard to piece together what, precisely, Pusztai's scientific blunder had been. But by August 14, one fact was unambiguous: after a career of distinguished service, with nearly three hundred scientific publications to his credit, Arpad Pusztai would be retiring at the end of the year. He was immediately relieved of all responsibilities for UK or European Union studies of genetically modified foods. In addition, he was ordered not to talk with the media about the results of his genetically modified potato studies.

In light of these dire personal and professional consequences, I later asked Dr. Pusztai if he would make the same statement again on television.

"Oh, yes. I did before the Parliament," he answered without hesitation. "For me it was a question of conscience. You see, we were the only people who were doing such studies, not just in Europe but in the whole world. I found most people were getting a bit worried about this. In the United States, by the end of 1996, they introduced genetically modified soya and genetically modified corn into the human diet by the back door. We didn't know anything about it," he continued.

"By the beginning of 1998, we were beginning to really get worried, because we did see this major interference with the development of some organs and the immune system, and none of these testing techniques that we used had been used by the companies. What I thought at the time—and I still think so, in fact, even more—was they have allowed something untested, potentially unpredictable, and potentially harmful to come into the food chain for reasons we could not really understand," he said.

"Genetically modified soya and corn are not better than regular soya and corn, so we really didn't have any benefit out of it. It is not cheaper. I put it at the time, and I still maintain, that in fact the companies decided they will be testing them on us."

Clare Watson of Genetic Concern! issued a press release stating that the conclusions of Pusztai's preliminary rat-feeding study were not the primary issue of concern for Irish consumers. After all, no potatoes on the market in Ireland contained genes from the snowdrop lectin. The only genetically engineered potato in the global food stream was Monsanto's NewLeaf potato. NewLeaf potatoes, which were marketed by a Monsanto subsidiary called NatureMark, contained a gene taken from the soil bacterium *Bacillus thuringiensis,* or Bt, rather than the snowdrop bulb lectin. What worried Irish citizens was the fact that none of the genetically modified soya and maize flowing into Ireland from the U.S. corn belt had been put through systematic feeding trials with rats and humans.

For its part, Monsanto welcomed the Rowett Research Institute's speedy disavowal of Pusztai's findings. "It is now clear that the alleged study did not involve genetically modified potatoes but used normal potatoes mixed with a well-known highly toxic compound," the company said. "There is no credible evidence that suggests that genetically modified food is unsafe."

When I called a scientist at the U.S. EPA's Office of Pesticide Programs and asked his opinion of Pusztai's study, he told me he had not heard about it. But he assured me that the genetically engineered Bt potatoes that the EPA approved for sale in 1995 were completely safe for human consumption. Agency scientists had reviewed studies submitted by Monsanto on its Bt potatoes and found them safe. In addition, the EPA had sought the advice of outside scientists, who also deemed the potatoes safe.

The potato study that ignited a firestorm of publicity in August 1998 throughout the UK failed to generate even a spark of interest in the U.S. media or among government regulators. I wondered what other kinds of genetically engineered food experiments were taking place, unreported and out of the public eye.

In early November 1998 I flew to Florida for a two-day conference on transgenic plants sponsored by International Business Communications, U.S.A. Several dozen biotechnology researchers, government officials, and businessmen gathered at the Buena Vista Resort in Disney World to hear a preview of some of the changes that biotechnology researchers have in store for the plant kingdom.

A Monsanto official gave the conference participants a peek at some of the products in the company's biotech pipeline. The biotech giant was looking ahead to the demographics expected in coming years and trying to anticipate some of the illnesses that might afflict people, he said, adding that biotechnology would provide "solutions for tomorrow's world." Population growth, the increasing wealth of the planet, and the aging of the population were the three factors driving Monsanto's push to develop biotech crops, he said. He talked about the possibility of developing crops with new "output traits," such as corn with different kinds of oil and chickens with amino acids that would make them yellow.

In the biotechnology industry, a distinction is often drawn between "input traits" and "output traits." Input traits include agronomic characteristics that have been engineered into plant seeds that are of interest mostly to farmers—disease resistance, herbicide tolerance, and the ability to kill insect pests. They are often intended to replace some of the chemical insecticides and fungicides that conventional farmers apply to crops. To the extent that they actually reduce the use of chemical pesticides, biotech crops can be viewed as beneficial for the environment. Bt cotton, for instance, has been shown to reduce the number of applications of pesticides that farmers must make to their crops each growing season.

The primary benefits of crops engineered with input traits accrue to growers and to the chemical company that owns patents on the seeds. When a farmer "buys" biotech seeds from Monsanto, for instance, he or she is actually purchasing a license to grow the seeds. Monsanto owns the patent on the seeds and restricts how the farmer is allowed to use them. A key restriction under the licensing agreements forbids farmers to save seeds from their harvest to plant the following season. The advantage for Monsanto and other biotech companies is that farmers must come back each year to buy—or more accurately, to lease the use of—new seeds. This practice disrupts farmers' age-old custom of saving seeds for future plantings and swapping seeds with neighbors. Biotech companies argue that without such licensing agreements, they would be unable to recoup the enormous investments that have gone into developing genetically engineered seeds.

Output traits, by contrast, are characteristics engineered into biotech crops that are intended to benefit food processors or consumers. Unlike input traits, which are not supposed to substantially change the nutritional profile of food plants, output traits are expected to add something of market value to the engineered crop. Biotechnology executives predict that foods with novel output traits will command premium prices because of their desirable qualities. Optimum, a high–oleic acid soybean oil engineered by Optimum Quality Grains, a division of DuPont, was among the first such commercial products. Promotional literature on Optimum claims that the new oil has an oleic acid content of 80 percent, compared with 24 percent for regular soybeans. The oil is also 33 percent lower in saturated fat and contains no trans fatty acids. The company says that these changes make high-oleic soybean oil a healthier cooking oil than olive oil, heavy-duty shortening, or regular soybean oil. Saturated fats and trans fat have been linked to elevated cholesterol levels, which studies have associated with increased heart attack risk.

Output traits are considered key to the "second wave" of genetically engineered crops. The "first wave"—canola, soybeans, and corn resistant to herbicides; cotton, corn, and potatoes that make their own pesticides; and a few disease-resistant fruits and vegetables such as papayas and squash—first began entering the market in 1996. In the coming "second wave," genetic engineers intend to make changes in the nutritional composition of foods. In theory, foods of the future will be "enhanced" to contain higher levels of certain beneficial nutrients than their garden-variety counterparts.

To this end, researchers at universities and private companies are investigating a class of nutrients known as *phytochemicals*. Many of these compounds have been linked to health benefits such as retarding the growth of cancer cells. Some scientists are probing the composition of tomatoes, which are high in a phytochemical called lycopene, and soybeans, which are high in phytoestrogens—substances thought to play a role in lowering the risk of cancer. Others are exploring the role that phytochemicals such as carotenoids, flavonoids, isoflavones, and folic acid might play in reducing heart disease and lowering cancer risk. Biotech companies are betting that the payoff from this kind of research could be enormous. The second wave of gene-altered crops possessing curative properties, however, has yet to come to fruition.

Ralph Hardy gave the conference participants a glimpse into a future world in which plants will be genetically manipulated to produce industrial products. Hardy, one of the nation's early pioneers in genetic engineering, and former DuPont official, is president of the National Agricultural Biotechnology Council at the Boyce Thompson Institute for Plant Research in Ithaca, New York. In the coming decades, he foresees fields of gene-altered plants producing oils, lubricants, vitamins, food additives, drugs, polymers, cosmetics, vaccines, paper, building materials, shipping and packing materials, household and personal care products, and cleaners. While economic growth in the twentieth century was based on nonrenewable fossil fuels, Hardy believes the twenty-first-century economy will be biobased. With substantial financial investment in agricultural research and development, he said, the United States could produce half of its energy and 90 percent of its chemicals from plants by 2050.

Hardy is not alone in his belief that plants will soon be genetically engineered to produce chemical and industrial products. Deans and provosts from twenty-two major agricultural research universities have signed on to his blueprint for the future of the plant kingdom. The scheme is spelled out in a brief document called "Vision for Agricultural Research and Development in the Twenty-first Century," which states that agricultural research and development into biobased technologies will improve the nutritional quality and safety of foods in the coming years. "Food will be modified to be more healthful with, for example, improved levels of antioxidants and balance of oil types. Transgenic plants and animals will produce health-related products such as pharmaceuticals and vaccines."

The "Vision" document acknowledges that the unchecked use of petrochemicals and destructive agricultural practices in the last decades of the twentieth century caused major environmental problems. "Fossil-based products, both in their manufacture and use, contaminate our air, water, and soil resulting in numerous environmental and health concerns." According to Hardy and his academic associates, the same chemical companies that wreaked havoc on the environment in the last half of the twentieth century will reverse course in the twenty-first, improving air and water quality and public health. They plan to accomplish this laudable goal by converting plants, trees, and animals into living factories.

It is hard to sort fact from fantasy in the publicity that is now pouring out of biotech companies and agricultural research universities. By constantly holding out a future full of fantastic benefits—fruits that won't rot after being picked; plants that will fix more nitrogen and thrive during droughts; nutrition-packed foods that will prevent cancer and heart disease for the world's wealthy; and high-yield, vitamin-enriched rice and cassava for the developing world's poor—the biotech enterprise keeps Wall Street investors optimistic. This guarantees a stream of capital into the laboratories of private companies and science departments of universities.

Martin Dickman, professor of plant pathology at the University of Nebraska, focused the attention of the conference participants on one specific experiment. Dickman's field is comparative pathobiology, which means that he looks for the common threads in disease development among plants, animals, and people. He himself has engineered

mammal and insect genes into plants in an effort to combat plant diseases. Dickman projected four pictures of mice onto a screen at the
front of the dimly lit meeting room. A large, disfiguring tumor pro-
truded from the thorax of a mouse on the far right. "We found that
a fungal ras gene with a single amino acid change causes tumors in
mice," he said.

The point of Dickman's study was to show that only a minor genetic change was needed to turn a gene from a fungus that attacks alfalfa into an oncogene—a gene that causes cancer in people. He later
gave me a more complete explanation of how he accomplished this
eerie feat. "We cloned a gene, that of an alfalfa pathogen. This gene is
a very well understood gene in humans, a ras gene," he explained. A
ras gene is involved in controlling the growth and development of a
cell. The ras gene in humans is one of the most commonly found mutations in cancer. In some cancers, it is actually the most common mutation, so scientists have studied the gene exhaustively.

"We cloned a ras gene from a fungus that attacks alfalfa," Dickman
continued. "The ras gene is about seventy-five percent similar to the
most common ras human gene. We took this fungal alfalfa pathogen
ras gene and made a change in one amino acid—one base, one nu-
cleotide. We changed one base pair out of eight hundred with a subtle
mutation."

When Dickman and his colleagues put the gene into living mice that
had been bred to be susceptible to cancer, they got tumors immediately. The ras gene from an alfalfa fungus with a single amino acid
changed had become "a bona-fide cancer-causing oncogene," he exclaimed, "even though this fungus has absolutely nothing to do with
cancer or tumor formation in its real life!"

"Is there any way that this type of gene could get out or affect people?" I asked him.

"Well, I suppose it could," he responded, adding that "it would be
difficult to be harmed by the gene we're working with because the fungus in the field doesn't have it, and this fungus never leaves the laboratory, so you'd almost have to ingest the fungus while you're in
the laboratory. It would be very difficult. We're not frivolous about it,
but I mean, it would be very difficult." He told me that after the experiments were finished, the fungi were all autoclaved. Researchers

commonly destroy the genetically engineered materials produced in laboratory experiments.

Martin Dickman seemed reasonable enough: he hoped to find ways, ultimately, to engineer disease-fighting capabilities into plants. But his experiment left me wondering whether any limits are placed on the kinds of genetically engineered organisms that university researchers are allowed to create in their labs.

Philip Regal, professor of ecology, evolution, and behavior at the University of Minnesota, has faulted the academic culture in which genetic researchers are trained for failing to instill an appreciation for unexpected changes and risks that might arise from their experiments. About a year after I attended the Florida conference on transgenic plants, Regal raised his concerns about the attitude of university research departments during an FDA meeting on genetically engineered foods. "It's really clear to me, as a university scientist, when I talk with genetic engineers, when I talk with molecular biologists, they cannot talk knowledgably about the risks," he told FDA officials. "Their textbooks tell them how to build these things, but they don't have chapters about safety that are meaningful. They're very, very thin."

Regal contends that genetic engineers have "a tendency to try to minimize the risks, and to try to deal with the incredible problems that genetic engineering presents with slogans and simplifications." He recounted his experience working some years ago as a consultant for a company that was experimenting with putting genes for spider venom and scorpion venom into corn, potatoes, and soybeans. "I think I talked them out of it. But you know, they had really convinced themselves that just because these venoms were not toxic to mammals, they didn't present any special problems," he said. "This rhetoric had worked at their minds, and they spent several million dollars doing this, and it was probably a very big mistake."

Some genetic engineering experiments, including Martin Dickman's work on the alfalfa fungus gene, are confined to laboratories. But many start in enclosed greenhouses and then graduate to open field tests. The power to grant or deny permits to researchers for outdoor tests of altered plants rests with the Department of Agriculture. The USDA has granted more than five thousand permits for field studies of genetically engineered plants and trees at more than twenty thousand locations.

I contacted Sally McCammon, a science adviser in the USDA's Animal and Plant Health Inspection Service, to find out what kinds of gene-altered plants researchers had grown in open "test" plots. She summarized the most common types of changes that researchers have been trying to make in food plants. In general, researchers want to slow down ripening or change nutritional characteristics of foods. Some are performing experiments to change the amount of sugar or the color in a fruit. Others are engineering food crops to produce vaccines or proteins for therapeutic use. Proposals to USDA for field tests of foods that have been genetically altered to make medical compounds generally contain confidential business information, and are not available for public review.

Many researchers are genetically engineering trees. Some are working on ways to change the metabolic pathways in trees to increase their uptake of carbon dioxide. Others are trying to reduce the amount of lignin in trees. Lignin is what makes woody plants tough. Sally McCammon suggested that reducing the amount of lignin in trees could prove beneficial because it might allow manufacturers to cut the amount of sulfurs and toxic wastes produced in the process of extracting pulp to make paper.

Genetic engineers at many universities and corporations have been experimenting with ways to "enhance" photosynthesis. Photosynthesis is what makes all life possible on Earth. Plants, bacteria, and algae are unique in their ability to turn energy from the sun into the food on which the rest of the animals and humans on the planet depend. "If we look back over the whole history of life on the planet, the evolution of photosynthesis by early bacteria was the most important single event," author Ernest Callenbach has written in *Ecology: A Pocket Guide*. "Without photosynthesis, Earth would have remained a dead planet."

Some of the changes that researchers envision for plants, such as making them more drought tolerant or salt tolerant, would offer important advantages to farmers in regions of the world where climate and soil conditions now make farming difficult or impossible. Other alterations in crops are being attempted strictly for the convenience of the food processing industry, which is always looking for ways to delay ripening, prevent bruising and discoloration, extend shelf life, alter taste and texture, and change processing and storage characteristics of food ingredients.

Monsanto tops the USDA's lists of applicants for outdoor releases of gene-altered plants, followed by Pioneer, AgrEvo, DuPont, Dekalb, Calgene (which is now part of Monsanto), and the USDA's Agricultural Research Service.

Employees with the USDA's Animal and Plant Health Inspection Service (APHIS) are supposed to oversee field tests to make sure that no "viable material" from thousands of experimental gene-altered plants and trees is left in the fields to propagate, and that no seeds and pollen are carried from the site by insects, birds, or rodents. Critics of government oversight of biotech field experiments have long questioned whether APHIS has had adequate procedures in place to make certain that seeds and "volunteer" plants are completely eliminated from test sites. In one of the earliest analyses of APHIS procedures, Tufts University researchers Roger Wrubel and Sheldon Krimskey found that the USDA has not required genetic engineers to provide information on the cold-hardiness of seeds or the length of time seeds from genetically altered plants can remain viable in the soil.

Wrubel and Krimskey studied twenty-eight of the environmental assessments that the USDA issued in response to university and industry researchers seeking permits to field-test transgenic plants between 1988 and 1990. "None of the environmental assessments addressed the potential movement of seeds out of the test site" by birds, rodents, ants, or beetles, the researchers wrote in an article in *BioScience* in 1992. APHIS personnel acknowledged that "it is possible that some seeds have escaped during field trials due to animal dispersal."

It is also possible that pollen has been carried from some of the experimental gene-altered plants, grown at thousands of test sites around the United States, to nearby wild plants. Some ecologists believe that the isolation distances used for experimental plants "may grossly underestimate the probability of long-distance pollen dispersal," Wrubel and Krimskey wrote. They noted that while one environmental assessment for genetically engineered squash assumed that four hundred meters was an adequate isolation distance, experimental work by scientists had found hybrids that had been produced between wild squash and cultivated squash separated by thirteen hundred meters.

There is no shortage of scientists queuing up to chop, transform, and rearrange the plant kingdom's ancient genetic inheritance. While researchers may sincerely hope to bring about beneficial changes in the plant kingdom, the question of pressing importance to the rest of us is: Who has the power to stop experiments that might inadvertently mangle the intricately woven and irreplaceable web of life?

FOUR

WHO'S MINDING
THE GARDEN?

T hree large federal agencies are entrusted with protecting the
American people and the environment from risks that might
arise from genetically engineered plants, foods, animals, and
microbes. Under a regulatory scheme devised during President Ronald
Reagan's administration, the Food and Drug Administration, the Environmental Protection Agency, and the Department of Agriculture
share the task of overseeing the biotechnology industry. In theory, that
sounds like a lot of firepower aimed at genetically modified organisms. A look back at the genesis of this triumvirate, however, reveals
that it was cobbled together more to reassure the public that someone
was watching over industry and university scientists than actually to
provide robust oversight.

The FDA has overall responsibility for safeguarding the American
food supply. It is charged with making sure that a new gene-altered
food is safe and nutritious for humans and livestock to eat, and that it
contains no new toxins or allergens. The USDA is supposed to ensure
that genetically altered plants do not threaten traditional agriculture
by spreading undesirable genes to native plants and weeds surrounding the farmlands. The EPA is responsible for protecting the public,
wildlife, and the environment from harmful man-made substances.
The agency's jurisdiction extends to genetically engineered foods when
a plant or microorganism has been engineered to make a pesticide or
other toxic substance.

At the time this regulatory framework was devised in the mid-1980s, genetic engineering was a brand-new technology, just over a decade old. In 1973 Herbert Boyer, a researcher at the University of California in San Francisco, and Stanley Cohen, a geneticist at Stanford University, performed a feat that stunned the scientific world: they transferred a gene from one organism to another, thereby creating an entirely new organism. As Eric Grace describes the experiment in *Biotechnology Unzipped,* they first used an enzyme to cut up large bacterial plasmids that contained a gene for resistance to an antibiotic. Plasmids are simple, circular structures found inside bacteria. Next, they cut up the DNA from an African clawed toad. They then mixed the fragments of toad and bacteria genes together, allowing time for them to recombine. When they added living bacteria cells to the mix, they found that DNA from the toad was taken in and copied in the bacteria.

The technique that Boyer and Cohen developed became the "classic" recipe for engineering bacteria to produce important medical substances and food processing enzymes. Scientists select and cut a gene with known characteristics from any organism and insert it into a bacterium, often *Escherichia coli,* or *E. coli.* The gene-altered bacteria are placed in large stainless-steel vats and allowed to multiply, producing millions of copies that synthesize the inserted gene product.

One of the first commercially successful products to be generated by bacteria, in so-called living "factories," was a synthetic human insulin, for use by diabetics. Bacteria have also been genetically engineered to produce food processing enzymes that the industry uses to control the texture, flavor, and appearance of many products.

As scientists began to broach natural barriers between species, moving genes from toads into bacteria, chickens into potatoes, and humans into pigs, some members of the general public became alarmed. Intuitively, many people felt that it was unethical or morally wrong to tamper with the natural order of life in so profound a way. Moreover, regardless of their moral or ethical position on the subject of genetic engineering, common sense alone told laypersons that scientists needed to exercise great caution in their laboratories, lest they inadvertently create synthetic organisms capable of causing catastrophic damage to other living organisms.

Scientists, too, realized that genetically engineered organisms could potentially cause irreparable harm to humans or the environment. In 1975, two years after the Boyer and Cohen experiment, molecular biologists took a voluntary time-out from their research to discuss its awesome implications. Biologists convened a meeting in Asilomar, California, a gathering now known as the Asilomar Conference, to discuss the potential hazards of genetic engineering. The National Institutes of Health organized a Recombinant DNA Advisory Committee (RAC) to oversee medical experiments in biotechnology. The RAC had jurisdiction only over research funded by the National Institutes of Health, however, and wielded no power over privately funded genetic experiments.

Almost nothing was known about how man-made organisms might behave outside the laboratory. Some members of the public, especially people living near major research universities, feared that exotic genetically engineered organisms might be released prematurely from scientists' laboratories. Others worried that scientists would accidentally create a deadly strain of virus. The 1971 hit movie *The Andromeda Strain,* based on Michael Crichton's science fiction novel about a deadly mutating pathogen entering Earth's atmosphere from outer space, had dramatized such fears. Beginning in the mid-1970s, many citizens' groups and some state governments began urging Congress to pass strong national legislation to protect communities from the release of potentially dangerous genetically engineered organisms. In 1976 the Cambridge, Massachusetts, City Council created the Cambridge Experimental Review Board to find out what kinds of experiments scientists were considering and to explore ways to protect the community from ill-conceived research.

The review board concluded that citizens cannot trust scientists to police themselves. "Knowledge, whether for its own sake or for its potential benefits to humankind, cannot serve as a justification for introducing risks to the public unless an informed citizenry is willing to accept those risks. Decisions regarding the appropriate course between the risks and benefits of a potentially dangerous scientific enquiry must not be adjudicated within the inner circles of the scientific establishment," board members wrote in a report to the city council. The city then formed a Biohazards Safety Committee, made up of or-

dinary citizens as well as public health officials. Cambridge went on to
pass an ordinance stipulating the conditions that scientists conducting
recombinant DNA research at institutions within the city would have
to meet.

As other state and local governments considered passing laws
and ordinances to regulate genetic research, the EPA stepped into
the vacuum in Washington and assumed the right to oversee the development of genetically engineered microorganisms. In 1983 Stanley
Abramson, then an attorney in the EPA's Office of General Counsel,
wrote a memorandum asserting that genetically engineered organisms
fell under the scope of the Toxic Substances Control Act. TSCA (pronounced "*toss*-ka") is the federal law that gives the EPA responsibility
for protecting the public and the environment from noxious chemicals.

Opponents of the EPA's stand argued strenuously that organisms
are living things, not chemicals, and should not have to undergo the
kinds of safety tests that the agency requires chemical manufacturers
to perform on new synthetic chemicals. Tough rules and expensive
testing requirements might squelch the development of genetically
engineered products, biotechnology advocates feared. But industry
leaders also recognized that some degree of federal regulation had its
benefits, in reassuring the public of the safety of new products and in
providing a predictable business environment. Before a company can
embark on new product development—especially when the product is
something as controversial as a new form of life—it has to have at
least a rough idea of how much it will have to spend on both research
and development and safety testing.

The EPA held a hearing in March 1983 to give interested parties a
chance to voice their views on the kinds of regulations that should
apply to commercial products being developed by the nascent biotechnology industry. Harvey Price, executive director of the newly
formed Industrial Biotechnology Association, spoke at the hearing.
"This last year was in fact the first year that a product for practical
purposes saw the marketplace," he told EPA officials. "It was the first
approval of a major product in the U.S. by the FDA, a type of genetically created insulin. This is the beginning of a very important period,
because it takes biotechnology out of the pure science era, where companies were very much concentrating on their research programs and

dreaming dreams and playing with the mice and other interesting phenomena, into the more economic period, where it's clear that biotechnology is not a gold mine but rather a business. It's driven by economic considerations to a significant extent and the other commercial factors that go in the marketplace."

With actual products to be made and real money on the line, the issue of biotechnology regulation heated up. Albert Gore, Jr., who was then a Democratic congressman from Tennessee and chairman of the Investigations and Oversight Subcommittee of the House Committee on Science and Technology, held the first congressional hearing on the environmental implications of genetic engineering. Gore opened the hearing on June 22, 1983, with a story about a French scientist named Leopold Trouvelot. In the late 1860s, the Frenchman brought into the United States a few eggs of an obscure variety of European moth, to try to breed a disease-resistant form of silkworm. While he was living in Medford, Massachusetts, moths hatched from the eggs, and a few of them escaped. These gypsy moths dug in, multiplied, and became one of the most destructive environmental pests in the country, defoliating ten million acres of trees in 1981 alone.

The proliferation of the gypsy moth, Gore said, "illustrates the problems that can occur when new foreign organisms are introduced into our environment. In the mid-1970s, as recombinant DNA technology became widely utilized in laboratory experiments, great public concern arose over the possibility that some deadly new virus or chemical kudzu might escape from a lab and endanger surrounding populations."

Gore warned, "The organisms that are being created today through genetic engineering are far more exotic than any creatures that have existed before. These are organisms with completely new genotypes, and, consequently, their potential for environmental damage could be far greater than for any 'natural' organism." Taking an unabashedly pro-environmentalist stand, he added, "It is important, therefore, that we understand all the potential environmental ramifications of an organism before it is released into an ecosystem—instead of waiting and finding out about them after the damage has occurred."

Martin Alexander, an agronomy professor at Cornell University in Ithaca, New York, reminded Gore's subcommittee that "a single species eliminated most chestnuts from fifty million acres" across America and a fungus spread rapidly through the U.S. corn crop in the 1970s, cutting yield by 10 percent. Sounding a cautionary note about the release of genetically engineered organisms, he said: "It is thus my view that alien organisms that are inadvertently or deliberately introduced into natural environments may survive, that they may grow, they may find a susceptible host or other environment, and they may do harm. The probability of all these events occurring is small, but the consequences of this low-probability event may be enormous."

The congressmen on Gore's subcommittee concluded that federal agencies had little ability to evaluate the likelihood of such a risk from releasing altered organisms outdoors. They did not call for the creation of a new agency, however; nor did they recommend new legislation to protect the public from genetically modified organisms.

Gore did continue to urge a go-slow approach to releases of genetically engineered organisms. "When people are confronted with the visions of new products and miraculous discoveries coming from new technology, they often wonder what is on the other side of the coin," he said at a public policy meeting on biotechnology at the Brookings Institution in 1985. "I believe that biotechnology has more potential to reshape the world as we know it than any other technology besides nuclear power. This time, public policy implications should be handled a little bit better than the implications of technology have been handled in the past."

On the other side of the debate, many university and industry researchers downplayed the potential risks of genetic engineering. Scientists were worried that a fearful public would shut down their research labs. Entrepreneurs contended that too much regulation of the fledgling field would stifle its potential and drive up the costs of developing new biotech products. Business executives argued that America would have to be willing to take certain risks in order to maintain its intellectual lead in this important new field. Restrictive regulations would place U.S. biotechnology companies at a competitive disadvantage, they insisted, and hence give European and Japanese biotech firms a technological edge.

Ralph Hardy, who has been influential in shaping U.S. biotechnology policy from its inception, questioned the need for strict oversight of an industry that had no history of mishaps. "The issue of government regulation to protect human health and the environment is perhaps of the most immediate concern to industry," he wrote in 1985. "If excessive, regulation can cause significant delays in product development, at a cost that can be catastrophic for small companies." He added, "It is worth noting that the proposed regulation of the biotechnology industry is unique. If regulations are imposed, this would be one of the few cases in which an industry has been made subject to significant health and safety controls before any hazards have been proved or any industrial accidents have occurred."

Hardy, who became director of life sciences for the DuPont company in 1978, had helped convince DuPont that it should move aggressively into the life sciences. "We were ahead of Monsanto," he told me. In 1981 the company speculated it would be 35 percent life science–based within a decade. "We hired a superb group of people. We couldn't build facilities fast enough. We actually bought some laboratories so we could have facilities for people," he recollected. But in July 1981 the ambitious plan was derailed when the company acquired a major stake in Conoco Oil. "With the acquisition of Conoco, there was considerable debt involved. The ability to move forward in life sciences at DuPont plateaued in 1981. I left in 1984, when I was fifty." He went on to become president of BioTechnica International, a biotechnology firm.

In 1985 American citizens who were aware of the new field of genetic engineering showed little willingness to allow scientists to tinker with plants and animals. To persuade the government to take a gamble on the new industry, some biotechnology advocates made the counterintuitive argument that it was actually riskier for society to be cautious than to take risks. Avoidance of risks can be harmful to society in the long run, they argued, because without a degree of risk-taking, beneficial discoveries might never be made. "Society cannot insist on zero risk. Society has to be willing to accept some risk in order to gain information that will provide the basis for greater certainty about risk and safety," David Jackson of Genex Corporation argued at the Brookings Institution forum in 1985. "The scientific community should be getting on with the job of identifying and performing the key experiments relating to the large-scale release of genetically modified organisms, so that regulatory policy having a sound, factual base will be possible."

Throughout the 1980s the eager biotechnology industry prodded the government to take a chance on new products developed through genetic engineering and the potential benefits they might bring society. The lightning rod for public debate over the risks and benefits of biotechnology was an infamous bacterium known as *ice-minus*. Steven Lindow, a researcher at the University of California at Berkeley, made the discovery that a specific protein produced by the bacterium *Pseudomonas syringae* is what causes ice crystals to form on plants. Lindow and his colleagues genetically modified the bacterium to eliminate the gene that causes the formation of ice, thus creating a so-called ice-minus strain.

Advanced Genetic Sciences (AGS), a California-based company, licensed the ice-minus technology from the University of California. AGS hoped to make a product to spray on crops that would considerably lower the temperature at which frost would begin to form on plants—an obvious boon to growers. Because the university received federal funding, Lindow sought approval from the Recombinant DNA Advisory Committee of the National Institutes of Health to test the effectiveness of ice-minus on plants in the field. In April 1983, following two RAC meetings, the NIH committee concluded that "since the organism was familiar and similar to bacteria that occur in nature, the field test presented insignificant environmental risks."

When the NIH approved Lindow's field test in June 1983, Jeremy Rifkin filed a lawsuit in the U.S. District Court for the District of Columbia seeking to halt the release. Rifkin, an author and director of the Foundation on Economic Trends, sought a thorough assessment of the possible consequences to the environment from the release of ice-minus bacteria, known as Frostban. A federal judge blocked the release, ruling that a detailed environmental assessment of the experiment was indeed required under the National Environmental Policy Act (NEPA). Then in October 1984, at the behest of President Reagan's Working Group on Biotechnology, the EPA announced that companies needed to receive an experimental use permit from the agency—the EPA—for small-scale releases of microbial pesticides. This move circumvented the NEPA requirement, because the approval of a release of a substance under an environmental law automatically satisfies NEPA. The EPA granted an experimental use permit to AGS in November 1985. But four months later, after learning that AGS had illegally tested ice-minus on fruit trees growing on the roof of the AGS facility, the EPA suspended the permit.

In April 1987 the EPA finally approved the AGS request to perform open-air tests of the ice-minus bacteria. The company made plans to spray Frostban on a plot of strawberry plants in a rural county in California. The first legal deliberate release of a genetically engineered bacterium was an event charged with tension, attracting reporters from around the world. The night before the test was to go forward, antibiotechnology activists entered the field and disturbed many of the plants. The release occurred as planned, and photographs of the historic event show a woman wearing a protective white moon suit and respirator applying the genetically engineered bacteria to a small patch of strawberry plants. The image, which was reproduced widely in the media, was not one to inspire confidence among a public that was already edgy about genetic engineering.

A few days after the application of Frostban to the strawberry plants, the EPA allowed AGS to perform another open-air test. Steven Lindow conducted the second test on a small potato plot in California—this time without donning a moon suit. While his experiment was more low key, attracting less media attention, vandals also attacked the potato crop.

The paper trail of government documents makes clear that in the 1970s and 1980s concerned Americans felt visceral fears and ethical reservations about genetic engineering. In Cambridge and California, residents who insisted that biotechnology experiments be performed indoors, in sealed containment facilities, were voicing many of the same worries as European citizens in the late 1990s. When U.S. citizens became aware of specific genetic engineering experiments that scientists were planning to launch, many demanded that the tests be curtailed—or at the very least closely monitored.

In the 1980s the EPA demanded extensive laboratory studies and information from biotech companies before it would grant permits for field tests. AGS, which had initially projected sales of $300 million annually from Frostban, ended up selling out to DNA Plant Technology. Frostban's new owner, too, ultimately decided against pursuing products that involved genetic engineering. Faced with a fearful public and cautious government regulators, many U.S. biotechnology companies scrapped their own plans for outdoor tests. University researchers complained that regulatory costs and uncertainties were preventing them from performing field tests of promising bioengineered plants.

Many scientists complained that zealous government regulation of genetically engineered organisms had served to reinforce the public perception that there must be something to fear from biotechnology.

In September 1987 Gary Strobel, a plant pathologist at Montana State University in Bozeman, grabbed national headlines when he performed an unauthorized outdoor field test on elm trees. He injected fourteen elm trees on the university campus with genetically altered bacteria in hopes of protecting them from the fungus that causes Dutch elm disease. Strobel had called EPA officials in June, seeking permission to begin his experiment before July. When permission was not forthcoming, he began the experiment in any case, worried that if the authorization came too late, he would have to delay the test for a full year. Following a reprimand from the EPA, Strobel decided to terminate the outdoor test. The tearful scientist was pictured on the front page of *The New York Times* on September 4, 1987, cutting down his experimental elm trees.

Despite these setbacks, the biotechnology enterprise ultimately found a sympathetic ear in Ronald Reagan's anti–big government administration. In 1984 the president's Domestic Policy Council formed a work group to develop a policy that would regulate the new lifeforms that scientists were creating in their laboratories. The group argued that comprehensive new legislation was unnecessary. Existing agencies and laws could handle safety reviews of the products of biotechnology, the White House concluded, based on the assumption that genetically engineered foods, drugs, and animals are inherently no riskier than their conventional counterparts. The administration issued a draft regulatory framework in late 1984.

Critics of the draft framework contended that the nation's food, agriculture, and pesticide laws had been enacted between 1910 and 1957 and as such were not intended to handle the powerful new techniques being developed by genetic engineers to manipulate life-forms. Trying to shoehorn biotechnology rules into laws passed before genetic engineering existed, they claimed, would leave the public unprotected.

Proponents of the framework countered that the government should regulate only the final product that issued from a biotech company laboratory. A squash is still a squash, they argued, even with a few genes from bacteria and viruses added to its genetic code. "Regulate the products—not the process of biotechnology!" became the rallying cry of the young biotech industry.

Biotechnology boosters, whose ranks included scientists and administrators at the nation's land grant research universities, several professional science societies, drug companies, medical research firms, major chemical companies, and food companies, insisted there was no evidence that the process of genetic engineering would introduce any unique problems to crops and other products.

In fact, in 1986 no track record of scientific experiments was available to back up this claim. Researchers had very little knowledge about how engineered microorganisms would behave in the natural ecosystem; nearly all of their experience was in genetically engineering bacteria under clean, controlled laboratory conditions. Scientists knew even less about genetically engineered food crops. Ralph Hardy had acknowledged in 1983, at Congressman Gore's investigations and oversight subcommittee hearing, that scientists had virtually no experience with genetically engineered plants. They had become quite skilled at using the technique of recombinant DNA to manipulate bacteria within laboratory and industrial settings. They could genetically engineer bacteria with a fair degree of precision to make proteins reliably. But they had little experience in transplanting novel microbes out of doors, where they would have to compete for their survival with indigenous microorganisms under harsh environmental conditions.

The first whole genetically engineered plant, a petunia, was not even made until 1983, so biotechnology boosters had no way to know how such plants would perform in an agricultural setting. "At this stage, we maybe know about twenty-five of the hundred thousand or so genes of any one plant. Within the last year a foreign gene has been transferred into a plant and made to function," said Ralph Hardy, who was then director of life sciences research at DuPont. "This is certainly a milestone event. It is not a practically useful one."

But industry advocates prevailed, and in June 1986 the White House Office of Science and Technology Policy published the final document known as the "Coordinated Framework for the Regulation of Products of Biotechnology." A key premise of the Coordinated Framework—which still provides the basis for U.S. oversight of gene-altered life-forms—is that plants and animals created through genetic engineering techniques are no different, either in kind or in nature, from the plants and animals that result from traditional methods of selective breeding. The biotech industry viewed the Coordinated Framework as a triumph, in large measure because it preempted congressional efforts to pass a law that might have required tough government oversight of genetically engineered products. A strong federal law could have slowed the development of genetically engineered foods, drugs, and animals by mandating stringent reviews and expensive batteries of safety tests.

But victory with policymakers in Washington solved only one of the biotech industry's major problems. The second big obstacle to the development of genetically engineered products was harder to manipulate: public perceptions. Mindful of most people's wariness of anything genetically engineered, biotechnology leaders in government and industry realized they had to convince the public that gene-altered organisms were neither foreign nor fearsome. Language became a key tool in their strategy to reshape the image of genetic engineering. Scientists who had proudly extolled genetic engineering as a revolutionary technology capable of creating entirely new organisms toned down their rhetoric.

Instead, science policymakers recast genetic engineering as an extension of the age-old practice of plant and animal breeding. They began to describe the process of cloning animal and bacteria genes for insertion into food plants as akin to the use of yeast by ancient civilizations to make bread. In its "time line of biotechnology," the Biotechnology Industry Organization dates the origin of biotechnology to between 4000 and 2000 B.C., with the use of yeast by Egyptians to leaven bread and ferment beer and the production of cheese and wine in Sumeria, China, and Egypt.

"Much of the mistrust of biotechnology stems from a disconnect between benefits of the so-called classical methods of plant and animal breeding, which are widely accepted as for the public good, and the molecular methods of breeding, which are not widely accepted as for the public good," R. James Cook, chief scientist for the National Research Initiative Competitive Grants Program at the USDA, told participants at the National Agricultural Biotechnology Council (NABC) meeting in 1994. The NABC, based in Ithaca, New York, began holding forums in 1989 to provide diverse stakeholders with the chance to discuss social, ethical, and policy issues surrounding the development of biotech foods. "Scientists—myself included—have perpetuated this disconnect by playing up the new biotechnology as 'powerful' and 'different' while not emphasizing enough until recently the continuum, interdependence and common goals of molecular and classical methods of breeding."

Cook's suggestion that "molecular" methods of breeding are comparable to "classical" methods blurred the distinction between a relatively new, sophisticated laboratory procedure and normal reproduction. To breed plants, according to the dictionary, is "to produce plants by selective pollination." Molecules don't breed—intact organisms breed. By definition, plant breeding, like animal breeding, requires sexual compatibility. But plant genetic engineering involves forcibly inserting genetic material from an entirely unrelated organism, such as a bacterium, another species of plant, or even an animal, into the genetic information of the plant—thereby achieving a transformation that could not, under any normal set of circumstances, result from the act of breeding. Breeding comes into play only *after* a series of sophisticated steps in laboratory conditions has transformed the plant.

George Wald, a professor of biology at Harvard University and a 1967 Nobel laureate, made it clear that there are no similarities to be drawn between classical breeding and genetic engineering. Recombinant DNA technology, he wrote, "places in human hands the capacity to redesign living organisms, the products of some three billion years of evolution. Such intervention must not be confused with previous intrusions upon the natural order of living organisms; animal and plant breeding, for example; or the artificial induction of mutations, as with X-rays. All such earlier procedures worked within single or closely related species. The nub of the new technology is to move genes back and forth, not only across species lines, but across any boundaries that now divide living organisms."

Wald further warned, "Up to now, living organisms have evolved very slowly, and new forms have had plenty of time to settle in. Now whole proteins will be transposed overnight into wholly new associations. Going ahead in this direction may be not only unwise, but dangerous. Potentially, it could breed new animal and plant diseases, new sources of cancer, novel epidemics."

Liebe Cavalieri, a molecular biologist at the State University of New York at Purchase, has also drawn a "clear distinction between classical breeding and genetic engineering." The insertion of virus, bacteria, and animal genes into plants "can raise the level of bioactivators in the plants to a point where they're not safe to eat," Cavalieri warned in 1998. Scientists who assert that genetic engineering is the same as classical breeding are making a "political statement enshrouded in loose science."

David Schmidt, senior vice president of the International Food Information Council in Washington, D.C., tracks the public's attitude toward genetically engineered foods, on behalf of the food industry. Through polls and surveys, he has found that people consistently respond negatively to the terms *genetically modified, genetically engineered, genetically altered,* and *novel foods.* Schmidt has advised biotech industry and government officials to avoid straightforward scientific terminology, especially the word *transgenic,* which evokes a powerful negative reaction in many people. Consumers have more positive reactions to phrases like *grown a better way* and *enhanced farming,* which avoid the issue of biotechnology altogether, and, if the word *genetic* must be used, to the term *genetically enhanced.*

In 1987 the National Academy of Sciences (NAS) released a report that bolstered the idea that potential risks introduced by genetic engineering are no different in kind from risks introduced by selectively crossing two food plants. The NAS is an august body of scientists that is often called upon to advise the government on complex or controversial science policies. The academy assembled a group of five scientists to review the state of the art of genetic engineering and to decide whether the release of genetically engineered microbes, plants, and animals into the environment could present society with any unknown risks.

"No evidence based on laboratory observations indicates that unique hazards attend the transfer of genes between unrelated organisms. Furthermore, there is no evidence that a gene will convert a benign organism to a hazardous one simply because the gene came from an unrelated species," the scientists stated in an influential, slender report entitled *Introduction of Recombinant DNA–Engineered Organisms into the Environment: Key Issues*. "Many thousands of distant genetic transfers have been carried out with R-DNA techniques, and the organisms with the new genes have the predicted properties: they behave like the parent organism, but exhibit the new trait or traits expected to be associated with the introduced gene or genes." Arthur Kelman of the University of Wisconsin in Madison chaired the NAS panel, which also included Wyatt Anderson, University of Georgia in Athens; Stanley Falkow, Stanford University; Nina Fedoroff, Carnegie Institution of Washington; and Simon Levin, Cornell University.

Some of the sweeping generalizations that the 1987 NAS panel made would prove overly simplistic in the coming decade, as scientists and farmers gained experience with gene-altered plants. In fact, another group of scientists convened by the NAS in 1999 to review the underpinnings of U.S. regulation of biotech food plants described many ways in which the process of genetic engineering could potentially give rise to harmful effects. "Introduction of transgenes into plants typically involves random integration of DNA into the nuclear genome," the 1999 NAS panel found. The new, modified plant "might have unexpected traits" or "unintended consequences."

The later panel's findings, however, came too late to have any effect on the development of the gene-altered corn, potatoes, soybeans, canola, and other foods that the federal government approved in the mid-1990s. The 1987 NAS document had provided the scientific rationale for relaxed federal oversight that allowed genetically modified food crops to begin entering the American food supply with the 1996 harvest.

In 1990 the Bush administration issued a document entitled "Four Principles of Regulatory Review for Biotechnology" that fleshed out the Reagan administration's 1986 Coordinated Framework with a set of regulatory "principles" aimed at smoothing the way for the speedy development of biotech products. The first principle stated: "Federal government regulatory oversight should focus on the characteristics and risks of the biotechnology product—not the process by which it was created." This federal directive meant that the FDA, the EPA, and the USDA scientists were not supposed to single out genetically engineered foods for review simply because they had been developed using a potent new technology.

This principle struck many critics of genetically engineered food as little more than an empty slogan masquerading as science. It flies in the face of the U.S. patent system, which has ruled that genetically engineered organisms are unique creations for which patent protection can be granted. It is also a position that has isolated the United States from other developed nations: Australia, Austria, Belgium, Denmark, England, France, Germany, Greece, Ireland, Italy, Japan, Korea, the Netherlands, New Zealand, Norway, Portugal, Spain, Sweden, and Switzerland have all elected to treat food plants produced by genetic engineering techniques as different from plants derived through conventional breeding. In those countries, genetically modified food plants are subject to different rules—and to strict labeling requirements—precisely because they are genetically engineered.

Many people find it hard to fathom the logic behind the assumption that the process of genetic engineering poses no unique risks—or that any problems inherent in the process will be revealed by a superficial examination of the end product. Yet the bias against looking for potential problems with the *process* of genetic engineering so thoroughly permeated the U.S. government and scientific establishment that some of the simplest questions about gene-altered foods remain not just unanswered but unasked.

The second of the four principles laid out by the Bush administration called on federal agencies to conduct speedy reviews and to ease the regulatory "burden" on industry. This principle said, "For biotechnology products that require review, regulatory review should be designed to minimize regulatory burden while assuring protection of public health and welfare."

The third principle stated, in part, that any regulatory programs "should be designed to accommodate the rapid advance in biotechnology." This principle enshrined Bush's belief that the expected benefits of biotechnology justified relaxed—if not outright lax—government oversight. The fourth and final principle stated: "In order to create opportunities for the application of innovative new biotechnology products, all regulations in environmental and health areas—whether or not they address biotechnology—should use performance standards rather than specifying rigid controls or specific designs for compliance."

In February 1991, several months after the principles were issued, Vice President Dan Quayle, who chaired the White House Council on Competitiveness, submitted a report to President Bush urging him "to oppose any efforts to create new or modify existing regulatory structures for biotechnology through legislation." Quayle also warned the president, "Foreign nations may seek to ban certain biotechnology products developed in the U.S. without sound scientific basis. They may also attempt to create informal barriers such as labeling and inspection requirements." Quayle's remarks indicated that government and industry leaders fully anticipated a strong negative reaction to genetically engineered foods abroad.

At the end of 1991, several biotech companies were getting ready to take their experimental foods into the marketplace. Calgene was engineering the Flavr Savr tomato. Monsanto was working on Roundup Ready soybeans and a potato engineered to make an insecticide. Calgene and Rhône-Poulenc had an herbicide-resistant cotton plant in the works, and Asgrow Seed Company was testing a virus-resistant squash. Having been bolstered by the four principles, the biotech industry was about to get another lift from President Bush.

On May 29, 1992, the Bush administration announced a new FDA policy on "foods derived from new plant varieties developed via recombinant-DNA technology." The policy stated that genetically engineered foods were to be considered no different from their conventional counterparts unless substantial changes had been made to the nutritional composition of the foods. Companies did not have to label genetically modified foods unless they had inserted a gene from a known allergen, such as a peanut, to the engineered variety or substantially changed its composition. The FDA did not require companies to perform premarket safety tests of genetically altered foods. Nor did they have to notify the FDA of their intention to market new genetically modified whole foods. Instead, companies were encouraged to have a voluntary "consultation" with the FDA. These consultations would take place behind closed doors.

The journal *Bio/Technology* reported that the biotech industry appeared to be "uniformly delighted," even "elated" by the FDA policy. Reporter Jeffrey Fox wrote that in 1992 Richard Godown, president of the Industrial Biotechnology Association in Washington, D.C., told him, "We can bring products to market without unnecessary regulations. But nobody has more incentives to bring safe and natural products to market than the companies." Another biotech industry representative, William Small, executive director of the Washington, D.C.–based Association of Biotechnology Companies, was quoted as saying he was "elated" with the new biotech food policy "because it treats 'food as food, regardless of the process used to produce it.'"

Many consumers who learned about the proposed policy sent the government angry letters. Of the five thousand citizens who submitted comments, 80 percent told the FDA that they wanted the government to require mandatory labels on genetically engineered foods.

Even the National Agricultural Biotechnology Council, a group friendly to genetic engineering, was critical of the Bush policy—though less for its substance than for the back-room politics behind it. The policy had been hammered out in Dan Quayle's Council on Competitiveness and foisted upon the public as a fait accompli. "Our feeling was that the FDA policy was fine, but the implementation was wrong," Ralph Hardy, president of the biotechnology council, later told me. "It was not a very transparent process. It didn't have the opportunity for public debate that it should have had." On the day the Bush administration announced the policy, Hardy recalled, the biotechnology council moved immediately to send letters to the secretary of health and human services, the secretary of agriculture, the EPA administrator, and Dan Quayle, assailing the process.

An internal memorandum from FDA commissioner David Kessler

to the secretary of health and human services, dated March 20, 1992, indicated that the biotech foods policy was hurried along at the behest of the biotechnology industry and the Council on Competitiveness.

"Companies are now ready to commercialize some of these improvements," Kessler wrote. "To do so, however, they need to know how their products will be regulated. This is critical not only to provide them with a predictable guide to government oversight, but also to help them win public acceptance of these new products." He added, "furthermore, the Biotechnology Working Group of the Council on Competitiveness wants us to issue a policy statement as soon as possible."

FDA commissioner Kessler's memo voiced unquestioning acceptance of the industry position that genetically modified foods would be beneficial for consumers. "The new technologies give producers powerful, precise tools to introduce improved traits in food crops, opening the door to improvements in foods that will benefit food growers, processors, and consumers," he wrote.

Kessler was prepared for an angry response from environmental groups. In the same memo, he wrote, "A coalition of environmental groups—the Environmental Defense Fund, the Natural Resources Defense Council and the National Wildlife Federation—wrote FDA last year urging that we routinely require formal food additive pre-market approval for foods from genetically modified plants and that we require that the fact that the food is from a genetically modified plant be disclosed in labeling." He predicted, "They may challenge our policy as leaving too much decision-making in the hands of industry and not adequately informing consumers."

Without a requirement that companies issue a pre-market notice of their intention to bring out a new biotech food, consumers would have no way of knowing when, or if, a particular food had been genetically engineered. FDA officials, like USDA and EPA bureaucrats, were in close contact with industry executives and knew what kinds of crops genetic engineers were designing. But concerned citizens and independent scientists could not to get hold of actual modified foods and test them because none were publicly available. When Kessler declared that biotech foods coming from the laboratories of chemical companies were "improved," Americans were forced to take his word on faith.

Unfortunately, the trust that citizens were asked to extend to FDA regulators did not stop there. FDA officials themselves were taking on faith industry's assurances about the safety and benefits of gene-altered foods. The 1992 policy gave industry a free hand to produce whatever genetically engineered foods it wished without pre-market approvals. At the time when the Bush administration developed the FDA policy, there wasn't a single genetically engineered seed on the market for farmers to plant, nor a single biotech vegetable for consumers to buy. Citizens didn't know if these novel food products would first reach the grocery stores in ten months, ten years, or far into the twenty-first century.

An impressive lineup of policymakers, from the vice president and the FDA commissioner to university scientists, insisted that genetic engineering would "enhance" foods for the benefit of consumers. In 1990 the American Medical Association (AMA) joined the list of trusted professionals promoting the benefits of these as-yet-nonexistent products. The AMA adopted a policy on genetically engineered crops that was based not on scientific health studies—none were available—but on the agricultural biotechnology industry's expectations and promises:

> It is the policy of the AMA to (1) endorse or implement programs that will convince the public and government officials that genetic manipulation is not inherently hazardous and that the health and economic benefits of recombinant DNA technology greatly exceed any risk posed to society; (2) where necessary, urge Congress and the federal regulatory agencies to develop appropriate guidelines which will not impede the progress of agricultural biotechnology, yet will ensure that adequate safety precautions are enforced; (3) encourage and assist state medical societies to coordinate programs which will educate physicians in recombinant DNA technology as it applies to public health, such that the physician may respond to patient query and concern; (4) encourage physicians, through their state medical societies, to be public spokespersons for those agricultural biotechnologies that will benefit public health; and (5) actively participate in the development of national programs to educate the public about the benefits of agricultural biotechnology.

The legal justification for the 1992 FDA policy turns on the statement that the agency "is not aware of any information showing that foods derived by these new methods (plant biotechnology) differ from other foods in any meaningful or uniform way, or that, as a class, foods developed by the new techniques present any different or greater safety concern than foods developed by traditional plant breeding."

But internal FDA documents made available to the public for the first time in 1999 reveal that many of the agency's own scientists believed otherwise. Several FDA scientists who were asked to review a draft of the proposed policy had advised James Maryanski that the process of genetic engineering *was* different from traditional breeding and that it might possibly introduce different kinds of risks into food plants. From 1996 to 1999, as billions of bushels of Bt corn and Roundup Ready soybeans poured into the American food chain, those professional opinions remained locked away in the FDA's files. The Alliance for Bio-Integrity's lawsuit against the FDA brought them to light.

Linda Kahl, an FDA compliance officer, had objected to the 1992 policy when it was first proposed, likening it to "trying to fit a square peg into a round hole. The first square peg in a round hole is that the document is trying to force an ultimate conclusion that there is no difference between foods modified by genetic engineering and foods modified through traditional breeding practices. This is because of the mandate to regulate the product, not the process," Kahl wrote in her January 1992 comments to Maryanski.

"The processes of genetic engineering and traditional breeding are different, and according to the technical experts in the agency, they lead to different risks." Furthermore, Kahl added, the acknowledgment "that the risks are different is lost in the attempt to hold to the doctrine that the product and not the process is regulated."

The second square peg in a round hole, Kahl said, involves asking scientific experts to generate the basis for a policy statement in the absence of any factual information. "It's no wonder that there are so many different opinions—it is an exercise in hypotheses forced on individuals whose jobs and training ordinarily deal with facts."

Louis Pribyl, another FDA scientist, rejected outright the notion that genetic engineering can be characterized as simply an extension of conventional breeding. "There is a profound difference between the types of unexpected effects from traditional breeding and genetic engineering which is just glanced over in this document," Priybl wrote in his comments to Maryanski.

FDA scientist Edwin Mathews warned that the levels of toxins in genetically modified foods could be higher than in normal foods. And Samuel Shibko told Maryanski that laboratory methods currently available to federal regulators would probably fail to find new toxins in genetically engineered foods.

FDA scientists had reason to worry that the insertion of foreign genes into a food plant might cause the plant either to produce higher levels of a toxic substance it already makes or to produce a new toxic substance in its edible portion. Naturalists and scientists have long known that food plants contain hundreds of uncharacterized substances, most of which imbue foods with their unique taste, smell, and texture. More than eight hundred "phenolic" substances, which contribute to the bitter taste and sometimes the color of foods, have been detected in plants. Phenolic substances include the tannins found in tea, coffee, cocoa, and grapes, as well as the flavones that give grapefruits and oranges their yellow color.

Many of the plant foods we commonly eat produce dangerous levels of toxins. Over centuries of trial and error, humans have learned to destroy harmful toxins by cooking them, or to avoid eating the poisonous part of a plant. Potato tubers, for example, contain toxins called solanines that, at high levels, can make people sick. Plant breeders in the United States and many other countries are required to test new varieties of potatoes to make sure the levels of solanines are not too high. Cyanide, found in peach pits and Asiatic varieties of lima beans, can kill people if eaten in high enough doses. A substance called myristicin, which is found in celery, nutmeg, parsley, dill, and black pepper, is also toxic at high levels.

Janet Andersen, who directs the Biopesticides Division of the EPA (which oversees biotech food plants), provided a compelling argument in support of the need for government regulation of new genetically engineered plants. In potato plants, the tubers, if properly cooked, provide a significant source of nutrition for a large part of humanity. But the leaves of the potato plant have natural pesticidal properties and contain a compound that can cause birth defects. If a genetic engineer were to move genes from the potato plant into spinach, in hopes of protecting the spinach plant against insects, the compound that causes birth defects might also be transferred to the spinach leaves, Andersen said during a congressional hearing. The nutritional composition of the genetically modified spinach leaves might remain substantially the same as normal spinach, yet the new plant would certainly raise a possible health risk to consumers.

In 1992 the FDA ignored the advice of many of its own scientists. President Clinton, following in Bush's footsteps, also chose to disregard the scientists' warnings. Even as questions about the adequacy of U.S. regulation of biotech foods mounted around the world in the mid- to late 1990s, Clinton administration officials continued to assert that genetically modified plants are substantially equivalent to their conventional counterparts.

In 1997 the FDA issued guidelines for the industry on what to expect from its private consultations on bioengineered foods. The guidelines acknowledge that the FDA neither requires nor reviews specific safety tests on genetically engineered foods: "During the consultation process, the FDA does not conduct a comprehensive scientific review of data generated by the developer."

James Maryanski has stated that ultimately it is the company developing a genetically engineered food, not the FDA, that is responsible for the food's safety. "Foods are not required to undergo pre-market approval by FDA. So new varieties of corn, for example, or soybeans, do not necessarily, do not come to FDA for approval before they go to market," he said at a public meeting. The Food, Drug and Cosmetic Act "places the legal responsibility for the safety of these products on the developer, on the purveyor of the product."

The FDA may believe that it is the job of biotech companies to make sure their own foods are safe. But a Monsanto spokesman told a writer for *The New York Times Magazine* that it is up to the FDA to ensure the safety of genetically engineered foods. "Monsanto should not have to vouchsafe the safety of biotech food," Monsanto's director of corporate communications told Michael Pollan, whose insightful article on genetically engineered potatoes was published in the October 25, 1998, issue. "Our interest is in selling as much of it as possible. Assuring its safety is the F.D.A.'s job."

FDA officials have argued that U.S. citizens can trust the biotech industry to thoroughly test the safety of genetically engineered foods in-house. Big brand-name companies have too much to lose by marketing dangerous products, agency thinking goes, to allow an unsafe food product on the market. But many consumers and activists in the United States and much of the rest of the world have found little reassurance in this line of reasoning.

When I had a chance to interview Maryanski, he told me the FDA was "very comfortable with the foods that are on the market." I asked him what he thought about the idea put forth by some scientists that genetically engineered foods might have the potential to cause immune system damage or other health problems to humans.

"People can postulate lots of things," Maryanski answered, "and it's not very helpful unless you have some information that you can actually look at to evaluate what's being said. You know, I think that we have looked at all of the genetic material that's been introduced into a plant from the perspective of what is the likelihood that this substance is likely to produce some adverse effect. Now, we can't guarantee the safety of any product that we regulate, whether it's a drug or a food or a food additive. The only thing we can do is make the best judgment based on the information that we have."

I raised the issue of labeling with him, reminding him that about 80 percent of the five thousand comments the FDA received on its policy in 1992 sought mandatory labels on genetically engineered foods. In subsequent surveys, between 70 and 90 percent of Americans polled have repeatedly said they want such labels. "It seems that's a pretty constant figure," I said. "Eighty percent of Americans want genetically engineered foods labeled as such. Why won't FDA do that for the public?"

"Because we first of all don't do things by vote," Maryanski retorted. "We do it based on the Food, Drug and Cosmetic Act, and the act describes the information that's required in labeling. And it really comes down to whether there's a consequence for the consumer, whether there's nutritional changes or some other substance in the foods that would have some consequence for the consumer. Does the consumer have to cook the food differently than they used to do, for example."

"Let me ask you this," I pressed. "Given that there weren't any long-term tests of these genetically modified foods on people or primates, if someone were to characterize this as really an experiment going on with the food, what would your response be? That U.S. citizens are engaged in an experiment to test the long-term safety—"

"In the same sense, with every other food, we know nothing about long-term effects of eating a particular food on any of our physiological systems," Maryanski responded. "If you're eating carrots over your lifetime, what do they do to your immune system or any other system? We don't know that."

FIVE

GUNPOWDER AND CORN EMBRYOS

T he U.S. government's contention that genetically engineered food plants are essentially the same as nature's own never rang true to critics of biotechnology. It began to sound hollow within the Clinton administration, when the White House bestowed national recognition on Monsanto researchers for their technical wizardry in developing genetically engineered crops. On December 8, 1998, President Clinton named four Monsanto scientists—Ernest Jaworski, Robert Fraley, Robert Horsch, and Stephen Rogers—as recipients of the prestigious National Medal of Technology. In a statement posted on the Internet, Clinton praised the researchers and their company for "pioneering achievements in plant biology and agricultural biotechnology, and for global leadership in the development and commercialization of genetically modified crops."

Ernest Jaworski, who headed the Monsanto research team, in the late 1970s had envisioned ways to use discoveries in the exploding field of molecular biology to create plants that did new and useful things. One of the useful things Monsanto's new plants could do was kill insects; another was to resist Roundup.

Jaworski identified three major technological challenges that Monsanto would need to overcome to succeed in its pioneering effort to engineer plants that make useful substances. The first hurdle was to identify that scientists could, in fact, put new genes into a plant. The second was to prove that the new gene would be expressed in the engineered plant. And the third was to establish that the gene expressed in the plant would be passed on to its progeny, to the next generation.

Jaworski, who describes himself as a lifelong learner, joined Monsanto in 1952, after earning a doctorate in biochemistry from Oregon State University. For a quarter of a century, he worked on agricultural chemicals, specializing in weed-killers. He worked on Ramrod and Vegedex and Randox, this last a chemical forerunner of Lasso, the herbicide that turned the company's struggling agricultural division around.

The 1950s and early 1960s were favorable times to be working in agricultural chemicals. The American petrochemical industry was booming, and the public was optimistic about the marvelous benefits that chemistry promised. Most citizens still viewed dichlorodiphenyl-trichloroethane (DDT) and other chemical pesticides as a boon to farm productivity, with no downside. In the decade leading up to the publication of Rachel Carson's *Silent Spring* in 1962, farmers, public health officials, and municipalities pursued an all-out war against insects. They pumped massive amounts of DDT, dieldrin, parathion, heptachlor, and other noxious pesticides into marshes and cranberry bogs and onto farm fields and roadsides. In her landmark book, Carson warned that these pesticides were killing birds and wildlife, poisoning waterways, and unleashing an epidemic of leukemia and cancer across the land. As the bald eagle approached extinction and one chemical-drenched waterway spontaneously caught fire, the harmful effects of synthetic chemicals became widely known.

The shine came off the chemical industry in the 1970s. On April 21, 1970, the first Earth Day, millions of demonstrators marched in the streets of American cities. These environmentalists ushered in a new way of thinking about science and technology and particularly about chemicals. Central to this shift in people's view of the Earth was a growing awareness that man-made pollution could alter natural forces on the planet and disrupt the web of life.

In 1971 thousands of gallons of dioxin waste from an herbicide factory spilled onto the roads of Times Beach, Missouri, leading to the evacuation of the town's entire population by 1973. Also in 1971 the United States banned the use of DDT. Insecticides and herbicides were found to be leaching from farmland soils into groundwater, running

off fields into lakes and rivers, and contaminating drinking water supplies with carcinogens. The EPA began to tighten its regulations on
agricultural chemicals.

It was against this backdrop that Ernest Jaworski turned to the
budding field of plant biotechnology. In the early 1970s he became interested in a new area of research called *plant cell tissue culture*. Scientists had learned how to grow plant cells in a "suspension culture,"
an aqueous medium in a shaker. "You could generate millions of plant
cells in a flask. The beauty of that is you could use plant cells much the
same way as you could use bacterial cells," he told me. Researchers
were able to grow bacteria in a suspension culture and then add various agents such as antibiotics to the flask. If scientists wanted to develop bacteria that were resistant to the effects of the antibiotic
kanamycin, for example, they could apply heavy loads of the drug to
the bacteria and then isolate the bacteria that survived.

Jaworski got the idea of using these techniques to see if he could
isolate and select out plant cells that survived application of the weed-
killer Roundup. Known generically as glyphosate, Roundup was an
important new herbicide: EPA scientists viewed it as safer than many
of the older compounds on the market. Roundup would eventually
become the world's top-selling herbicide. Jaworski did not develop
glyphosate, but he was one of the first scientists to study how it
worked. He investigated the biochemical pathways in plants that
make them susceptible to Roundup, which is a broad-spectrum herbicide capable of killing most green plants.

"Nobody was doing any work on the mode of action at that time,
but it intrigued me, because this was a chemical that killed all plants if
you sprayed the leaves, and yet it was not toxic to animals," he recalled. "So I reasoned that it had to be inhibiting some mechanism
unique to plants, not found in animals."

Beginning in the mid-1970s, Jaworski tried to interest the agricultural section of Monsanto in his idea for transforming crop plants to
make them able to tolerate Roundup. In 1979 he finally found support for his idea from a new executive, Howard Schneiderman, who
was keen to move Monsanto into biotechnology. Schneiderman suggested that Jaworski move from the agriculture section to the corporate research program, where he would have more time and latitude
to do exploratory research.

In 1980 Jaworski became Monsanto's director of biological sciences. He then assembled a team of scientists, including Robert Fraley, who was doing postdoctoral work at the University of California in San Francisco; Steve Rogers, an assistant professor and microbial geneticist at Indiana University; and Robert Horsch, a cell biologist who was doing plant cell tissue culture work at the University of Saskatchewan.

In 1981 Monsanto entered into a collaborative research agreement with Washington University in St. Louis to pursue biomedical research. That same year Monsanto sold its stake in its joint Conoco Oil venture to DuPont, signaling a move away from chemicals and into new technologies. The sale was among the many business decisions that would gradually transform Monsanto from a gritty chemical manufacturing company into a "life sciences" company. In 1982, following the acquisition of DeKalb Genetic's wheat research program, the company formed Monsanto Hybritech Seed International. A year later it acquired Jacob Hartz Seed Company, which specialized in soybeans. The following year Monsanto dedicated a new Chesterfield Life Sciences Research Center, which it located outside St. Louis. In 1985, in a major acquisition that led the company into pharmaceuticals, Monsanto purchased G.D. Searle, maker of the sweetener aspartame.

In 1993 Robert Shapiro was named president of Monsanto. A graduate of Harvard University and Columbia University Law School, Shapiro believed fervently that the future of plant protection and agriculture lay with biotechnology. He embarked on an aggressive mission of acquiring seed companies to obtain the germ plasm to which the company's patented genetic engineering techniques and products could be applied. Under his leadership Monsanto went on a buying spree, spending more than $6.5 billion to purchase seed companies and consolidate the company's dominance in plant genetic engineering.

In January 1998, when the company's life sciences metamorphosis was complete, Monsanto took out a full-page ad in *The Washington Post*, heralding its new image and mission. "At 97, we have a new outlook on life. A better life for our planet," the ad copy said. The company's new slogan, printed in an elegant italic font floating on an empty white page, suggested an end to dreaded dioxins, PCBs, and toxic pesticides. Three words, printed in blue-green ink beside a child-like drawing of a simple vine, bore Monsanto's altruistic new motto: "Food Health Hope." The multinational corporation was bowing out of pollution and usurping the mantle of life-giver.

Many of Monsanto's scientific breakthroughs in designing profitable new products from food plants grew out of research carried out at universities and industry laboratories around the world. Genetic researchers had devised two basic methods for inserting foreign genes into a plant. The first technique is truly ingenious. It involves smuggling the desired foreign genes into the DNA of plant cells by attaching them to a tiny mobile structure that dwells inside the soil bacterium called *Agrobacterium tumefaciens*. It is basically the same way many a frustrated dog owner manages to get a pill down the gullet of unwilling pet. Most dogs do not want to take a pill, any more than a plant cell's genome wants to take in a piece of DNA from another organism. Just as the dog will taste the bitter pill and spit it out, the plant's genetic structure, trained over millions of years of evolution to maintain its integrity, will often inactivate or break down the foreign DNA. But if the pet owner stuffs the pill deep inside a piece of frankfurter, the dog eagerly swallows the food, carrying the pill into its gut.

Similarly, genetic engineers figured out a way to slip foreign genes into plant cells by taking advantage of a technique devised long ago by *Agrobacterium,* a bacterium that causes crown gall disease in susceptible plants. Crown galls are disfiguring, tumorlike growths that infect fruit trees, grapevines, and other plants. In 1975 two Flemish researchers, Jeff Schell and Marc Van Montagu, discovered that the tumor-inducing genes of *Agrobacterium* are carried on a plasmid, or mobile unit of DNA, within the bacterium. Normally all of the bacterium's crucial hereditary information—its DNA—is found on a single chromosome. But this plasmid, Schell and Montagu discovered, is not part of that chromosome. Hence the bacterium could transfer the tumor-inducing genes carried on these plasmids into the chromosomes, or DNA, of the targeted plant.

Once the gene is inside the chromosome, its tumor-inducing properties kick in, causing the host plant's own cells to start dividing rapidly. The excessive cell growth results ultimately in the formation of the crown gall tumor. Researchers working independently at different laboratories in the late 1970s and early 1980s wondered if it might be possible to remove the tumor-causing gene from the crown gall plasmids and substitute genes that would produce something useful. Among those scientists were Jaworski and his colleagues at Monsanto.

But many major crop plants, such as corn, rice, soybeans, and wheat, are not susceptible to infection by the *Agrobacterium* method. Genetic engineers needed to find another way to insert foreign genes into wheat, corn, and rice—the three crops that make up more than half of humanity's diet. The next leap forward in the genetic alteration of food became possible with the invention of a powerful new tool, the gene gun. The device, also called a particle gun, was developed in the early 1980s by John Sanford and Theodore Klein, two Cornell University plant scientists at the New York State Agricultural Experimental Station in Geneva, New York.

The gene gun revolutionized the emerging technology of genetic engineering by giving scientists a formidable way to shoot foreign genes into organisms. Researchers could coat gold or tungsten particles one micron in diameter with thousands of genetically altered DNA molecules and then fire them into living cells and tissues. Hundreds, even thousands of shots would miss their mark or kill the cells. But against all odds, this improbable machine sometimes succeeded in delivering a new gene into the DNA of a living cell, forever transforming the organism's genome.

Scientists at Pioneer Hi-Bred International's gleaming Carver Center in Des Moines, Iowa, routinely use gene guns to create new crop varieties. Pioneer is the largest seed company in the United States, employing nine hundred researchers at 140 sites in twenty-nine countries and spending more than $130 million a year on research. I visited the Carver Center to see how researchers transform ordinary corn into varieties that kill insects or resist weed-killers.

"In the transformation process, the first thing you need to do is identify and isolate the gene. In the case of Bt corn, for example, the Bt gene was first isolated," Sam Wise, a field support trainer and tour guide for Pioneer, explained as we entered a suite of research stations.

Wise spoke in a steady, measured tone, finding simple words to explain complex material—a skill honed by years of guiding visitors from all walks of life through Pioneer's biotechnology labs. "Then you need to prepare a transformation vector. The transformation vector is going to include the promoter, the target gene, and the terminator." Scientists sometimes compare the transformation vector to a "cassette" containing the bundle of genes that they are attempting to load into the host plant's genome.

Wise noted that genes don't work independently. "They've got to be told to turn on and turn off. They may work with other genes," he explained. "In the case of the Bt gene, it gives you the desired protein that kills the European corn borer larvae. And also, since this is a very random process, we need to be able to find it when we put it into the plant." For this reason scientists add what is called a *bacterial selectable marker* gene to the cassette of genes inserted into the plant.

Scientists need the marker because inserting foreign genes into plants is a relatively haphazard process, with low rates of success. Markers are used in order to distinguish the rare individual plant into which the foreign gene was successfully incorporated from the many that were not successfully transformed. One early marker researchers used was a gene from fireflies that expresses luciferase, the enzyme responsible for making the firefly's tail glow. In genetic engineering experiments, plants into which the firefly gene has been incorporated light up, whereas plants without the luciferase gene do not. In the late 1980s scientists began to make common use of antibiotics as markers to aid them in finding those plants into which a desired gene has been incorporated.

Genetic engineers have routinely used the antibiotic drugs ampicillin and kanamycin to see which of their experimental plants have incorporated the genes bundled up in the transformation vector. Any transformed plants that have successfully incorporated the new genes into their own genomes can be easily spotted, because they continue to thrive when placed in a growth medium that includes the antibiotic drug. Plants that are not naturally resistant to the applied antibiotic and have not incorporated the new genes cannot resist the effects of the antibiotic, and die.

"Quite frankly, that's where some of the concern comes with some of our European friends, because they're concerned about this bacterial resistance, or resistance to ampicillin," Wise told me. Many European doctors have expressed alarm over the practice of routinely inserting genes that are resistant to valuable antibiotics into genetically engineered food crops. Their worry is based on the promiscuous nature of bacteria, which swap DNA at a rapid rate. Antibiotic resistance develops when bacteria change, or mutate, in a way that reduces or eliminates the effectiveness of antibiotic drugs. Overuse of antibiotics promotes the spread of antibiotic resistance. Already several important antibiotic compounds have become ineffective against infections that they once cured. Some strains of tuberculosis, for example, are now resistant to all available antibiotics, and experts have found that some food-borne illnesses caused by salmonella and campylobacter are on the rise. In fact, the Centers for Disease Control has called antibiotic resistance one of the world's most pressing public health problems.

Some doctors fear that the antibiotic resistance genes in genetically modified corn and soybeans might exchange DNA with the millions of bacteria that inhabit our digestive system and pass along resistance to antibiotic drugs. Pharmaceutical companies may be able to develop new antibiotics capable of killing bacteria that have acquired resistance to older drugs. But, European critics argue, why take the chance of exacerbating a difficult medical situation—especially for the purpose of developing genetically modified foods that are not needed to begin with?

Advisers to the UK Ministry of Agriculture warned in 1995 that a gene resistant to the antibiotic ampicillin in gene-altered corn made by Novartis was powerful enough to break down ampicillin in the human stomach within thirty minutes. An effective course of treatment with antibiotic drugs requires multiple days of effective doses of the drug. Doctors rely on ampicillin and related antibiotics such as penicillin and amoxycillin to treat bronchitis, pneumonia, salmonella, septicemia, and other serious infections.

For years U.S. federal regulators dismissed these objections, arguing that the likelihood of creating antibiotic-resistant bacteria through this practice was remote. The FDA endorsed Calgene's use of a gene for resistance to kanamycin when the agency reviewed the Flavr Savr tomato. In 1995 James Maryanski said that after evaluating the relevant scientific issues, the FDA had found that the kanamycin-resistance marker gene "would not affect the clinical effectiveness of kanamycin in people taking the drug orally."

But five years later U.S. scientists and doctors started to change their thinking. Sanford Miller, dean of the Graduate School of Biomedical Science at the University of Texas, acknowledged in 2000 that American scientists had made a mistake in using kanamycin- and ampicillin-resistance marker genes in food plants.

Ampicillin should not have been used, Miller told me, because it is a widely used human antibiotic. "See, the trouble with the whole industry has been—not really a trouble, but historically, the situation is, these companies were founded by scientists. They wanted to make a construct, and they made a construct. None of them were concerned about making Frankenfood," he continued.

"They never thought through what the implication was of that marker being transferred to another microorganism. They never thought about that. It's not the way they thought. Now, they do think about that," Miller added. "There are better markers they can use today that are not antibiotics." Scientists can use markers that fluoresce, or light up, for example, to see if a plant has taken up a new gene.

Sam Wise emphasized to me that at Pioneer Hi-Bred researchers remove the bacterial marker once the transformation vector is constructed. "Now you've got your transformation vector," he said. "So you just magnify that. You make a whole bunch of DNA, and then eventually you get enough of it so you can actually see it in a test tube."

As he spoke, we approached the gene gun. The device is a square box about the size of a large microwave oven, with two rectangular chambers stacked one on top of the other, separated by a narrow compartment. There is a tray at the bottom of the device on which to set a petri dish. A round pressure gauge sits on the front panel, which is made of glass, providing a clear view of the inside. Wise referred to the gene gun as "an antique," but it is the type of gun that was used to make many of the genetically engineered crops we eat today.

If we had been conducting an experiment, we would have put several corn embryos on a petri dish and then used the gene gun to bombard the embryos with DNA-coated tungsten particles. But for this demonstration, we dispensed with the embryos. Wise put a petri dish in the bottom half of the clear chamber. "This is our target," he explained. Then he put liquid tungsten beads in a vial and shook it vigorously. The gun "actually uses a twenty-two blank cartridge to propel this," he said, holding a bullet between his thumb and forefinger.

"Like the twenty-two you'd shoot on a rifle range?" I asked.

"Yes, exactly. Pretty precise science, right?" He chuckled. "I put this twenty-two caliber on there, and put the firing pin on there. We're not going to fire it yet, though. If we did, what would happen is you'd just blow that dish apart. So we'll put this stopping plate in." He showed me a thick, clear plastic disk. "The stopping plate will stop the plug and allow the tungsten beads to go along and then spread out." Next, he created a vacuum in the box. "The reason you want a vacuum is that those little beads don't need any wind resistance," he explained.

"Watch right there," he said, then depressed the button that fired the gun. I heard a faint *pop* and saw a tiny flash of orange light.

Wise then let some air back into the chamber and opened the door. He pointed to some big gray specks—about the diameter of a pinhead—in the petri dish, where the corn embryos would have been in a real experiment. "Those big holes would kill the cells, but those little bitty holes are what you're looking for," he said, indicating even smaller specks.

He handed me the stopping plate, which had become mangled during the shot. The acrid odor of gunpowder lingered on the plastic disk.

"There are about thirty-nine patents involved in that shot," he added, as we entered a room housing the newer gene guns, sleeker models that use compressed helium to shoot DNA-coated gold beads into corn embryos.

Genetic engineering is a hit-or-miss affair, and success rates with the gene gun technique—even using the newer guns—are exceedingly low. "You have a one percent, or less than one percent, success rate with this—and that just means you got something in there," Wise explained. "You may not have got what you wanted, or it may not have

attached." If the gene lands in the wrong position in the host plant's
DNA, it might disrupt the normal functions of the plant. We walked
slowly past rows of glass petri dishes lining the shelves in the growth
chamber. Some contained rubbery-looking white growths called cal-
lus. Others contained green tissue. Researchers have to transform
hundreds of embryos before they achieve a variety with the desired
trait. Even when they do, the new trait may be expressed in the transformed plant at the expense of other key crop qualities. The new plant
might yield less grain, for example, than the nonengineered parent
plant.

Gene guns and *Agrobacterium* gave scientists the tools they needed to
clear the first hurdle that Ernst Jaworski had articulated for the Monsanto research program: putting foreign genes into plants. The second
technical challenge—getting the inserted gene to fully and reliably express the intended trait in the engineered plant—proved extremely difficult to achieve. In fact, in 2001, two decades after Jaworski set out to
make gene-altered crops in his laboratory, it still presented researchers
with technical problems.

That scientists encountered difficulty in getting a plant to express
the trait coded for by a foreign gene inserted into its genome should
not have come as a great surprise, critics of genetic engineering point
out. We need only look at the specialized cells within our own bodies
to realize that the mere presence of a gene does not mean it is constantly expressing all of its possible traits. While every cell in our bodies contains the set of genetic instructions to make the proteins that
digest food and grow hair, for example, only the cells in our stomach
will signal the production of digestive enzymes, and only our hair fol-
licles will sprout new locks.

"No cell will ever make use of all the information coded in its
DNA. Cells divide the work up among one another—they specialize,"
said Ricarda Steinbrecher, a scientist with the Women's Environmental Network Trust in London. "Brain cells will not produce insulin,
liver cells will not produce saliva, nor will skin cells start producing
bone. If they did, our bodies would be chaos! The same is true for
plants: root cells will not produce the green chlorophyll, nor will
leaves produce pollen or nectar."

Steinbrecher, an opponent of genetic engineering, once conducted research on gene regulation and therapy at the Royal Free Hospital in London. But after learning that engineered food was being commercialized in the United States, she became disenchanted with the uses to which genetic engineering was being put. "This is still very much just a laboratory science," she told me. Like many critics of the technology, she believes that genetic engineering has been prematurely applied to food production, before the ramifications for safety have been thoroughly explored.

"All in all, gene regulation is very specific to the environment in which the cell finds itself and is also linked to the developmental stages of an organism," she said. The leaves of a poppy plant do not turn red, even though the gene that endows the poppy flower with its brilliant red-orange hue is present in the DNA of the leaf cells as well as the blossom cells. In addition, expression of some traits is age dependent. Adults do not generally grow a third set of teeth. And as scientists learned, a tomato plant will not necessarily be frost resistant even after scientists have inserted into its DNA a cold-resistance gene from a deep-ocean-dwelling fish.

Scientists therefore had to find a way to reliably "turn on" the action of the genes that they inserted into their genetically engineered plants. The use of viral promoters solved part of the problem. Viruses have the genetic instructions needed to replicate themselves, but they lack the biochemical material necessary to carry out the replication themselves. So viruses reproduce by integrating their own genetic material into the DNA of a host cell. The DNA of a virus has a powerful control element, called a promoter, that directs the host cell to multiply. Genetic engineers, by taking a promoter from a plant virus and sticking it in front of the information block of a transplanted gene, were able to "turn on" the action of the gene in its new host plant.

A viral promoter isn't properly a switch, however, because it cannot be turned off. The promoter gene is always turned to the "on" position. So in the case of a Bt corn plant engineered to make a pesticide protein, for example, the plant churns out the toxic protein around the clock—a situation not found in nature.

In genetic engineering, viral promoters generally perform as expected, by turning on the inserted gene—but not without glitches. Often quite unpredictably, a transplanted gene will function for a limited time in the transformed plant and then inexplicably stop working. Genetic engineers call this phenomenon *gene silencing*.

Jean Finnegan of Canberra, Australia, and David McElroy of the University of California at Berkeley studied the problem of gene silencing. To find out how widespread the phenomenon was, they surveyed scientists at thirty firms that were trying to commercialize genetically engineered crops. Almost all of the scientists said they had encountered some level of gene silencing in their experimental engineered plants.

The levels of expression of inserted genes were "unpredictable" and varied greatly among genetically altered plants, Finnegan and McElroy wrote in a 1994 article entitled "Transgene Inactivation: Plants Fight Back!" in the journal *Bio/Technology*. Often the transplanted gene was unstable even within a single growing season. Their findings suggested that farmers would not necessarily be able to depend on bioengineered seeds to produce safe, reliable crop yields.

Finnegan and McElroy warned that as more genetically modified crops "leave the controlled environment of the research greenhouse and are subjected to the natural variation in the farmer's field," we could expect to see more problems with gene silencing.

The gene silencing problems companies encountered in 1994 still plague researchers today. In 2000 I asked James Siedow, a biology professor at Duke University and past president of the American Society of Plant Physiologists, to explain the phenomenon of gene silencing.

"As near as anyone can tell, gene silencing is a mechanism that plants have evolved to try to deal with viruses. So they've come up with this mechanism that says, 'If suddenly there's an overexpression of a particular gene, it's a good idea to stop it,' " Siedow explained. He acknowledged that genetic engineers were surprised when they first discovered that plants had ways to turn off the effects of inserted genes.

"Gene silencing clearly involves multiple mechanisms, not all of which are very well understood," he continued, adding that "it's an issue that can be problematic."

I asked him about the widely publicized efforts of genetic engineers to make strawberries frost resistant by adding genes from ocean-dwelling fish. Articles in newspapers and magazines often cite the dazzling feat of transferring flounder genes into strawberry plants as an example of the power of biotechnology. Greenpeace activists sometimes wear "Fishberry" costumes or carry placards featuring a half-fish, half-strawberry creature, to exemplify how biotechnology breaches natural boundaries.

"That's not commercialized. That didn't work," Siedow said. "We keep getting beat over the head with that one, but it didn't work."

I asked him why scientists have not been able to get the fish gene to work in strawberries.

"It was a long shot to begin with." Fish and strawberries, he responded, "live in a different kind of environment."

The kinds of genetically engineered foods that appear in our grocery stores have more to do with what university researchers and chemical companies have been able to do—and profit from—than with what is desirable. Consumers might have been inclined to embrace genetically modified foods if the first products to issue from companies had been allergy-free peanuts or cancer-fighting fruits. What the world got, instead, were potatoes and corn that make pesticides.

In the late 1980s, as Monsanto and other biotechnology companies moved their experimental plants from laboratory dishes and greenhouses into field trials, two traits alone accounted for nearly all the genetically engineered food crops: herbicide resistance and Bt.

SIX

PESTICIDE IN A SPUD

Once scientists learned how to put foreign genes into plants, they began to search for candidates they might use to profitably alter traditional foods. In the early 1980s two scientists at the Washington University in St. Louis found a promising candidate: *Bacillus thuringiensis*. Bt is one of nature's most powerful insecticides. It was first discovered in 1901 by a Japanese scientist who identified the previously unknown bacterium as the cause of an outbreak of disease that was killing silkworms. In 1915 Ernst Berliner isolated the bacterium from a dead Mediterranean flour moth that he found in a grain mill in Thuringia, Germany. Scientists have identified some thirty *Bacillus thuringiensis* subspecies. Each strain of Bt makes its own unique toxic crystals. One Bt strain kills moths and butterflies, the lepidopterans; another kills flies and mosquitoes, the dipterans; and a third kills beetles, the coleopterans. In 1981 Helen Whitely and Ernest Schnepf of Washington University identified the gene in Bt that is responsible for killing insects. The discovery helped turn the world of crop protection inside out.

The EPA has approved more than 175 Bt microbial formulations as pesticides for use on fruits, vegetables, orchards, and ornamental plants. The prospect of inserting the Bt gene *inside* the DNA of a crop plant—instead of spraying a Bt pesticide on the outside of the plant—offered a potential new product and profit stream for chemical companies seeking to make a business from the life sciences.

The life cycle of Bt is at once a splendid and deadly affair. Ordinarily Bt exists simply as an oblong, rod-shaped bacterium, but when it becomes nutrient-deficient and is ready to die, it produces an oval-shaped spore and a crystal. *Kurstaki,* the most common Bt subspecies, produces a bipyramidal, eight-sided crystal. Embedded within the exquisite Bt *kurstaki* crystals are toxic proteins. When a farmer applies a natural Bt microbial spray to plants, both spores and crystals cover the leaves. A European corn borer eating the leaves of a corn plant will ingest both the spores and crystals.

"The crystals dissolve in the insect's gut, because the insect gut is very alkaline, it has a very high pH," Susan MacIntosh explained. MacIntosh is the product safety manager for biotechnology products at Aventis CropScience. A protein chemist, she worked at Monsanto as a researcher from 1987 to 1990 on the development of Bt plants.

The corn borer's gastric juices chop the Bt protein into small pieces, including a piece that scientists call the toxic fragment. "That's the piece that's in all the Bt corn today—that toxic fragment," she continued. The toxic fragment then attaches to a cell in the membrane of the insect's gut. How the insect actually dies is a matter of conjecture among scientists. Some say the gut contents leak into the insect's blood, which eventually leads to its death. Others believe the insect dies from a lack of nutrition, since it stops eating after ingesting the Bt toxin. In any event, within two or three days the insect dies.

Bt sprays are popular with organic farmers, who are legally prohibited from using synthetic pesticides on their crops. Because Bt crystals occur in nature, they have become a weapon of last resort for organic growers battling intractable infestations of insect pests such as the Colorado potato beetle.

Conventional growers often use Bt sprays in fruit orchards and on economically valuable vegetables such as artichokes and celery. As insecticides, however, Bt sprays present problems and benefits, depending on the grower's point of view. For one, ultraviolet rays from sunlight break down the active agents in a Bt spray within about forty-eight hours, so that the toxin does not persist for long on the leaves of plants, in the soil, or elsewhere in the environment. This is not a problem for organic growers, home gardeners, and even some conventional farmers, who tend to have small plots and can time their applications of Bt sprays based on the appearance of specific pests and favorable climatic conditions. The very properties that make Bt foliar sprays benign for the environment, however, make them impractical to use on large, industrial-style farms. Bt microbial sprays do not produce bug-free fields. Pests that feed on roots, inside stems, or on other parts of the plants that the Bt sprays can't reach continue to thrive.

In 1988 Monsanto requested permission from the USDA to conduct field tests of genetically engineered Bt tomato plants. The USDA granted the company permission, and Monsanto put the plants into the ground. But the tomato plants with the transplanted Bt genes failed to produce enough of the toxic protein to kill insects.

The reason the Bt gene did not work, as scientists were to discover, has to do in part with the changes that organisms have undergone through evolutionary time. While all organisms use the same basic genetic code—the strings of A, T, C, and G that instruct cells to make amino acids and proteins, the building blocks of life—the instructions play out differently in primitive organisms such as bacteria than in complex organisms such as plants.

In tinkering with the Bt gene, Monsanto scientists found that bacteria use different coding sequences, or *codons,* than plants normally use to make the same amino acid. As Ernest Jaworski told me, "It took a lot of good molecular biology to modify the gene so that it still coded for the correct protein, but it would be expressed at much higher levels in the plant, because now you were using codon sequences that the plant prefers."

I was puzzled by the notion that plants and bacteria "prefer" different coding sequences. "My understanding of the universal genetic code had always been that you have to have the exact same sequences of amino acids," I pressed.

"You do have to have the same amino acid sequence for the protein, in general—otherwise it may not function the way it's supposed to. But you don't have to use identical codons," Jaworski explained. "In the case of plants, they are different enough from bacteria that if you know the amino acid sequence for the bacterial gene, and you know the coding preferences that plants use, then you can substitute some of the coding sequences—the codons—for different amino acids," he continued. "It sort of becomes empirical, which ones you switch."

Jaworski's description of how genetic engineering actually works bears little resemblance to the simplified explanations that experts generally give laypersons. The process of moving a gene for a given trait from one organism to another is often compared to movie editing. In this analogy, a gene is "cut" from one organism, copied, and "pasted" into the genome of another organism, in much the same way that a scene in a movie is cut from one place on a long spool of tape and inserted somewhere else. "DNA and videotape are linear information systems that carry encoded information. That information can be decoded, expressed, copied, spliced, and edited. More copies—i.e., more plants—can be made from the edits," says an explanation of plant genetic engineering on Cornell University's website. "The new 'tape' is intellectual property; so gene sequences—like videos—can be licensed, patented, and sold. Where there is money to be made, there is commercially viable technology and license agreements to be negotiated."

This analogy may make sense as a way to visualize the classical recombinant DNA techniques used to make bacteria produce insulin. It may also accurately reflect the justifications put forward by those who support the patenting of life-forms. But the linear, tape-to-tape analogy falls far short of capturing the complexity involved in the genetic transformation of food plants. As single cell organisms without a nucleus, bacteria are separated from plants by billions of years of evolution. The process of genetically altering plants sounded to me more like trying to snip a scene from an old eight-millimeter strip of black-and-white celluloid and paste it into a color VHS tape. A clever technician might be able to physically insert a segment of celluloid into a VHS tape, but if you put this hybrid tape into your videocassette recorder and hit the play button, it wouldn't work. It might even jam up the whole machine.

Susan MacIntosh's work at Monsanto involved analyzing the level of Bt protein expression in the thousands of plant cells into which researchers had engineered the Bt gene. She described to me the efforts of two of Monsanto's top scientists, who set about making "wholesale changes all across the protein" in order to increase the amount of Bt toxin that gene-altered plants would produce. The scientists made "a synthetic gene and they put it into the plant, and the expression went up five-hundred-fold!" she exclaimed as she recollected the exact moment in the laboratory when researchers realized that the Bt expression problem had finally been solved. "It was unbelievable! It was amazing."

In 1990 Monsanto created the first successful Bt crop, a Bt cotton plant. Monsanto scientists also put Bt genes of the subspecies *tenebrionis* into Russet Burbank potatoes. They engineered the potatoes to make a Bt toxin called Cry3A that kills Colorado potato beetles.

In 1991 Monsanto asked the USDA for permission to grow the new potatoes in test plots in several regions of the country. The USDA approved the request, and the company began field tests to find out whether the new potato plants would kill the beetles in the field. The plants did the job. In fact, they delivered a much higher dose of toxin to the pests than external Bt microbial sprays deliver.

Once Monsanto proved that its Bt potatoes killed Colorado potato beetles, the company's next challenge was to convince federal regulators the genetically engineered product would be safe for people to eat and safe for wildlife, including pollinators such as bees and butterflies. In September 1994 Monsanto submitted a petition to the USDA asking the department to "deregulate" the company's experimental Bt potatoes. When the USDA deregulates a genetically engineered plant, it removes all restrictions on plantings of the crop, clearing the way for the company to sell the bioengineered seed to farmers. Monsanto told the USDA that field tests on the Bt potato had shown that the gene-altered lines were equivalent to the parental Russet Burbank potatoes. The USDA agreed with Monsanto and deregulated the Bt potato in March 1995.

Because Monsanto's NewLeaf potato plants had built-in pesticides, the EPA had to decide whether they were safe for humans and for the environment. The EPA was accustomed to requests from companies to apply all manner of noxious pesticides to food crops—but always externally. Some of the chemicals that were applied to fruits and vegetables could be washed off, while others degraded over time. The

EPA was also accustomed to making fine calculations about how much of a particular pesticide residue remaining in or on a food was too much. Fruits such as peaches, strawberries, and apples absorb more pesticides than do oranges and bananas, which have thicker, more protective outer skins. The EPA's Office of Pesticide Programs must decide how much of each of the hundreds of chemicals registered for use on crops people can ingest without becoming ill. The EPA calls the maximum legal pesticide residue allowed on each food a "tolerance."

But genetically engineered foods were truly something new. The idea of intentionally adding new pesticide proteins to food plants pushed the science of gene-altered foods into uncharted territory—too far, in the view of some critics. Still, if plant pesticides were new to the EPA, many of the companies applying to register these novel products were familiar. The agency had years of experience working with Ciba-Geigy (which later became Novartis), Monsanto, Dow Chemical, and AgrEvo (which later became Aventis) on their applications to register new chemical pesticides. So for Bt plant pesticides, the EPA went about regulating them by starting with what was familiar: the Bt toxin in pesticide sprays.

The EPA regulates pesticides under the Federal Insecticide, Fungicide and Rodenticide Act of 1947 (FIFRA, pronounced *fif*-ra). Whenever a company makes a claim that a new product kills bugs or pests, the product becomes subject to EPA review. FIFRA defines *pesticide* very broadly, as "any substance or mixture of substances intended for preventing, destroying, repelling or mitigating any pest." Using this definition, garlic and fish oil sprinkled on a plant to ward off unwanted bugs or rodents become pesticides. The EPA's Office of Pesticide Programs actually does classify certain natural substances as pesticides and has categorized them as "biopesticides." Because the EPA makes the assumption that biopesticides are more benign than other categories of pesticides, including chemical agents, it does not require companies to perform as many tests to prove their safety as it requires for a new synthetic compound.

Since the Bt gene was derived from a naturally occurring bacterium that had a long history of safe use in organic Bt sprays, the EPA decided to classify the toxic protein produced by NewLeaf potatoes as a

biopesticide. It dubbed the new Bt crops, including all of genetic material necessary to produce the pesticide effect—the Bt gene, the antibiotic resistance marker gene, the promoter and enhancers—as "plant
pesticides." The term was a fairly accurate description of this new creation—a literal pesticide-in-a-plant—but it rankled many biotechnology supporters. "Plant pesticides" sounded ominous, they claimed. In
nature, they noted, many plants produce substances that protect them
from insects, yet the EPA does not regulate them as pesticides. The
critics also raised the "product versus process" argument, charging
that in regulating Bt plants, the EPA was actually regulating the
"process" of genetic engineering by which the new plant was created.
The agency countered that some regulation of the process was justi-
fiable because it was possible for scientists to introduce potentially
dangerous substances into otherwise familiar food crops via genetic
engineering. To appease critics, however, it eventually changed the
name of Bt plants from "plant pesticides" to "plant incorporated pro-
tectants."

The EPA required Monsanto to perform tests to show that the
toxin in its Bt potatoes would not harm honeybee adults and larvae,
parasitic hymenoptera, ladybird beetles, green lacewings, and the
northern bobwhite. Monsanto also fed a high dose of the Bt pesticide
protein to mice to make sure it was not toxic.

When I went to the EPA's public docket in Arlington, Virginia, to
look at the documents Monsanto submitted to prove its Bt potatoes
were safe for people and wildlife to eat, I noticed something odd. The
diets of the test animals contained neither genetically modified potatoes nor Bt proteins extracted from the gene-altered potatoes. Instead,
test animals were fed a different substance, which Monsanto called a
"substitute Bt protein," produced by *E. coli* bacteria. Several of the
studies Monsanto submitted to the EPA were actually intended to
prove that the toxic protein produced by the *E. coli* bacteria is "substantially similar" to the actual toxin produced inside the NewLeaf
potatoes that the public would soon be eating.

The reason for the substitution, Monsanto officials claimed, was
that extracting the actual Bt toxin from the NewLeaf potatoes would
cost too much, because the toxin is produced at very low levels. So in
order to churn out the relatively large quantities of the Bt substance

needed to feed to test animals and insects, Monsanto said, it had to rely on using a "substantially similar" substitute. The EPA went along with this line of reasoning.

In fact, it became standard practice for the EPA to accept the biotechnology industry's use of substitute Bt proteins to test its new products for human safety and ecological effects. This practice has drawn criticism, first from some international scientists, and later, during the comprehensive review of Bt crops undertaken by the National Academy of Sciences (NAS) in 1999. The NAS scientists said that current studies indicate that "plant-expressed Bt proteins are probably without human health risk." Nevertheless, NAS argued that tests "should preferably be conducted with the protein as produced in the plant." In 2001 the industry practice of using substitute proteins became a critical point in the dispute over the presumed safety of Star-Link, the genetically engineered corn variety that sparked a national food crisis.

Arpad Pusztai contends that scientists cannot and should not assume that a toxin produced in *E. coli* bacteria is equivalent to the actual toxin made in a Bt potato. "The *E. coli* product is different from the one produced by the plant," he insists. "Just imagine, the *E. coli* and the plant are evolutionarily billions of years apart. So although the DNA may be the same, when it gets into the machinery of being read and translated, the end product will be different."

Even in plants as closely related as the pea and the kidney bean, Pusztai and his colleagues have found, the same gene produces a different effect. In an experiment, they took a gene from kidney beans and inserted it into peas. The gene they transplanted, called the *alpha amylase inhibitor,* interferes with the digestion of starch. When they fed the pea plant to test animals, the expected problem with starch digestion disappeared within five minutes in the rats' gut.

"So the pea had treated the product differently from the bean, and these are both legumes," Pusztai explained. For this reason, he said, it is important for scientists to test each gene-altered food case by case.

In their study on the effects of genetically modified potatoes on rats, Pusztai and his colleagues did not rely on a substitute protein made by *E. coli.* "What we did was actually isolate the snowdrop bulb lectin from the potatoes. The companies always say, 'Oh, you can't do this, because the expression level is too low.' With the Bt toxin, for instance, they say the expression level is too low. But the expression level is the same as we had. In our case, it was between twenty and twenty-five micrograms per gram potatoes. That level is roughly the same as any other genetically modified product." He added, "I think they are either lazy or they were just covering up."

The EPA allowed Monsanto to make other substitutions in its Bt potato safety testing as well. One test required the company to demonstrate that the genetically engineered proteins would break down quickly in the human stomach. Monsanto submitted test data showing that the proteins "degraded within 30 seconds in simulated gastric fluid." There was no way to judge how closely this "simulated gastric fluid" matched conditions in actual human stomachs.

The EPA did not require Monsanto to conduct any long-term feeding studies of actual genetically modified potatoes on mammals, primates, or humans. Long-term testing was not needed, it reasoned, because organic farmers have used preparations of Bt microbial sprays on vegetables for decades without any reports of damage to human health. Critics pointed out that there is a world of difference between the limited exposure that wildlife have to Bt sprays and the exposure to Bt potatoes with a built-in Bt toxin. Organic farmers apply Bt sprays to potato plant leaves for only a few days each growing season, whereas the Bt potato is genetically engineered to churn out the toxic protein all season long. EPA reviewers apparently did not view this difference as important.

The EPA also had to decide if the new proteins would behave in the same way as normal dietary proteins. The agency made the assumption that all of the genetic material inserted into the NewLeaf potatoes along with the Bt gene—the antibiotic resistance genes, *Agrobacterium*, viral promoters and enhancers, and other DNA fragments—would pose no health effects to people or to wildlife. In its "product characterization" of the Bt potato, the EPA wrote: "DNA is common to all forms of plant and animal life and the Agency knows of no instance where nucleic acids have been associated with toxic effects related to their consumption as a component of food."

Arpad Pusztai and his colleagues in the UK took a different approach in their study of potatoes genetically engineered to express the snowdrop lectin gene. Pusztai did not assume that anything about the DNA transformation of food plants is necessarily equivalent to natural foods. When he set out to develop a blueprint for safety assessments of gene-altered foods, he designed experiments to investigate the *process* of genetic engineering itself to see if it might produce unexpected adverse results. For this reason, instead of having just one control group—rats fed a nonengineered parent potato—against which to compare the rats fed the engineered potato, he also fed a third group of rats nonengineered potatoes mixed with purified snowdrop lectin. In his controversial study, he found that rats eating the engineered potatoes showed changes in organ weights and had gut lesions, but the rats ingesting regular potatoes to which the purified snowdrop lectins had been added did not.

Pusztai and other scientists then began to consider what factors in the genetic engineering process itself might account for the damage to the rats' guts. One suspected problem is the positioning of the inserted gene in the plant's DNA, which may have somehow disrupted the normal functioning of the potato plant. Another is the Cauliflower mosaic virus. Genetic engineers have used the Cauliflower mosaic virus promoter in virtually all of the genetically engineered foods on the market.

"All parts of the construct, including the Cauliflower mosaic virus promoter, are suspect because they've never been looked at," Pusztai told me. "This is a product-driven science—I hesitate to use the word *science,* because it is extremely sloppy science. We are putting in a lot of genes. It's not just a single gene.

"The Cauliflower mosaic virus 35S promoter is what we call a naked piece of DNA, which could be highly infectious," he continued. "We have no direct evidence for it," he was quick to add. "If you consider just for a moment, the Cauliflower mosaic virus 35S promoter is highly homologous to hepatitis B and HIV in the way it works."

"What makes U.S. scientists so certain it is safe, then?" I asked. When I had earlier questioned an EPA official about the virus, he responded that humans have a long history of exposure to the Cauliflower mosaic virus in their diets, with no known harmful effects.

"People are always given half-truths, and they are more dangerous than a full lie, because the lie can be exposed," Pusztai responded. "Cauliflower and broccoli can both be infected. When you look at cauliflower and you see these black spots—which people normally cut out, incidentally—that is part of the Cauliflower mosaic virus that has been attacking the cauliflower. So they say we've been eating this stuff for a long time, and it's quite true. Nobody can argue with it."

But when we eat the virus in its naturally occurring form, it is covered by a "coat" that surrounds the virus like a bowl, he explained. "Our immune system knows the 'coat' of the Cauliflower mosaic virus. We've been exposed to it, and it is the protein part, this coat part, which once it gets into our gut, our immune system will seize it under normal conditions—unless, of course, you are HIV positive or something, in which case you have no immune system—and then neutralize it. But the 35S promoter is part of the virus, and it is naked—there's no coat protein attached to it. Therefore, our immune system does not recognize it."

Mae-Wan Ho, a geneticist of the Open University in the UK and a vocal opponent of genetic engineering, has pointed out that the Cauliflower mosaic virus shares sequence similarities to human retroviruses such as the AIDS virus, human leukemia, and hepatitis B virus. She has raised a concern that when the naked Cauliflower mosaic virus 35S promoter is ingested in food, it might recombine with human viruses. The technology for genetically engineering food plants is so new that such theoretical problems have not yet been thoroughly explored, Ho asserts. Her scientific viewpoint has generated terrific controversy and has drawn scorn from many mainstream genetic engineers.

James Siedow at Duke University, for one, vigorously disagrees with Ho. "The vast majority of what she says is just incorrect," he told me. "For one thing, if there were problems with this promoter, it would have appeared years ago because cauliflower and even broccoli are loaded with 35S promoter, and we've been eating that. It's natural."

Arpad Pusztai remains unconvinced. "Again, it's really strange," he countered. "I can assure you there has been no publication even on taking the 35S promoter and putting it into the gut of a rat to see what happens. There's been no such experiment. Sometimes you say, 'Everything must have a few holes in it,' but here there are nothing but holes. There is no substance to it," he said of the technology. "I'm not a scaremonger. I'm not a doomsday merchant. I don't know what the effects are, but I would have thought that looking—with some basic knowledge of DNA—we ought to really have a look at what is happening."

Pusztai continued, "Look, we have never eaten 35S promoter without its coat. It may be totally harmless. What I'm saying is that it's worth testing it."

"Why do you think the testing hasn't been done?" I asked him.

"The reason why it is not done is because you will not get money for it. You will never get money for it. And if you get results, you will have considerable difficulty publishing it, because the biotech companies together with the politicians of the governments are hell-bent on pushing this technology on. All these studies cost a lot of money, and because all these GM crops have been patented, they want to get their money back as soon as possible. Even our small project on genetically modified potatoes cost 1.6 million pounds. And it was only a start. In the United States and in Canada, there's something like forty GM crops."

The U.S. FDA has a worldwide reputation for thorough, vigorous testing of new drugs. In fact, it was the FDA's outstanding track record with drug regulation, Pusztai said, that initially may have lulled European officials into assuming that the agency's approval of genetically engineered foods meant they had been rigorously tested.

Many scientists argue that toxicity tests capable of demonstrating the safety of foods simply do not exist. Humans know that conventional food sources are safe only because we have been eating them for millennia. As biotechnology advocates often point out, many of the foods now in grocery stores contain low levels of toxic substances that, if subjected to regulation today, would probably prevent them from being approved as safe for humans to eat.

To the assertion that biochemical tests are not available to determine the safety of either genetically altered or traditional foods, Pusztai responded, "That is simply not true. We can go through a testing procedure that will ferret out the major problems. It will not be a hundred percent guarantee, but as it is now, we have absolutely no guarantee." Researchers can design tests to find out whether a new bioengineered food has an effect on growth, development, the immune system, the endocrine system, the reproductive system, and other functions. "If you do the testing properly, it will show up any major problems." But Pusztai was quick to acknowledge there are limitations to testing: "We cannot do a number of tests. For example, we cannot use children and pregnant women in clinical trials."

Critics of Pusztai's rat study contend that problems with its experimental design make the results impossible to interpret. After extensively reviewing Pusztai's data and claims, the British Royal Society found them essentially useless. "Some results showed differences in the overall body weights and in the weights of individual organs in the two groups of rats. However, such results as were statistically significant did not fall into a readily discernible pattern," the British scientists wrote.

The Royal Society also charged that Pusztai "attempted to cover too much ground with the resources available." Reliably answering the pressing question of whether the genetic modification of food has harmful effects would require a major research effort, the scientists reviewing Pusztai's work said. "It would be necessary to carry out a large number of extremely complex tests on many different strains of GM and non-GM potatoes," the Royal Society wrote. By contrast, Pusztai's work concerned only one species of animal "fed with one particular product modified by the insertion of one particular gene by one particular method. However skillfully the experiments were done, it would be unjustifiable to draw from them general conclusions about whether genetically modified foods are harmful to human beings or not. Each GM food must be assessed individually."

But no independent researchers in the United States or the European Union were doing the kinds of complex safety tests on genetically modified foods that the British science establishment claimed to be so important.

Before it decided whether Monsanto's Bt potatoes would be safe for people to eat—and safe for wildlife and the environment—the EPA sought the advice of its Scientific Advisory Panel. This group of outside scientists advises the agency on difficult technical questions, particularly when the science has controversial public policy ramifications. The panel met to discuss Bt potatoes on March 1, 1995, in Arlington, Virginia. It was a historic meeting. The EPA was about to approve the first gene-altered pesticide-in-a-plant. Monsanto, Novartis, and AgrEvo all had applications pending before the agency for Bt corn and Bt cotton plants. Two weeks after its deliberations, on March 16, 1995, the panel delivered an eleven-page report to the EPA with a short—some would argue cursory—response. "The Monsanto B.t.-potato presents little potential for human toxicity," the scientists stated. Their conclusion was based on a review of studies performed by Monsanto.

In May 1995 the EPA approved Monsanto's genetically engineered Russet Burbank Bt potatoes for sale. Three months later, in August 1995, it approved the first application from a biotech company to license the sale of Bt corn. The Bt variety, called Event 176 Field Corn, was engineered by Novartis to produce the Cry1Ab protein. It produced relatively high levels of Bt toxin, as scientists would discover in 1999, after investigating the relative risks different brands of Bt corn posed to butterflies.

In October 1995 the EPA approved Monsanto's application to register Cry1Ac cotton, known as Bollgard Bt cotton, engineered to kill cotton bollworms and two other insects, the pink bollworm and the tobacco budworm. The product had to gain approval for human food consumption because cottonseed oil is used in foods. By December 1996, when the EPA approved a Monsanto request to register a variety of Bt corn called MON810, applications for genetically engineered plant pesticides were becoming fairly routine.

Then in April 1997 AgrEvo applied for a license to sell a genetically engineered Bt corn called StarLink, which produced a protein called Cry9C. The Cry9C gene comes from the Bt subspecies *tolworthi*. All of the other genetically engineered Bt corn varieties on the market contained genes from the Bt subspecies *kurstaki*.

EPA scientists immediately noticed something different about the Cry9C protein in StarLink corn: it is not readily digested in human gastric juices; nor does it break down in heat. These two characteristics are common to many of the food proteins that cause allergic reactions in susceptible people—from skin rashes to breathing problems to anaphylactic shock and death.

The possibility that genetic engineers will add new allergens into the human food supply is one of the thorniest issues facing the development of genetically engineered crops. In 1994 the FDA, the EPA, and the USDA held a joint meeting in Annapolis, Maryland, to address this problem. They invited some of the world's top allergists to talk about our ability to predict whether a given protein will cause an allergic response. Lacking a history of dietary exposure, the scientists concluded that there is simply no way to predict whether a new genetically engineered food has the potential to cause allergies.

Uncertainties notwithstanding, scientists advising the government in 1994 did come up with a framework to guide the fledgling biotechnology industry. They laid out a set of criteria to help genetic engineers identify the proteins that are most likely to cause allergies: Is the novel protein heat stable? Does it resist breakdown by enzymes and acids in the human stomach? Is the amino acid sequence of the protein similar to that of other known allergens? Is the novel protein relatively large? An answer of yes to any of those questions was considered suggestive, though not necessarily predictive, that the new protein might be a human allergen.

When EPA scientists considered applications to approve the three other genetically engineered Bt proteins—the Cry3A in Bt potatoes, the Cry1Ab in Event 176 corn, and the Cry1Ac in Bt cotton—they used the four criteria as a screen in an attempt to keep new allergens out of the food supply. When companies performed laboratory tests to simulate conditions found in a human stomach, they discovered that the three proteins all quickly degraded in gastric enzymes at low pH and broke down readily when they were subjected to heat.

But Cry9C protein in StarLink corn failed two of the four screens: it did not break down in simulated gastric juices and withstood heat tests at ninety degrees centigrade.

Viewing the Cry9C protein as a potential allergen, the EPA rejected AgrEvo's request to license StarLink Bt corn.

But AgrEvo was eager to break into the burgeoning Bt corn market, which was dominated by Monsanto and Novartis. The company decided to seek a license for limited use of StarLink corn for animal feed and industrial use. In May 1998 the EPA granted the company's request, approving the planting of StarLink corn on a limited acreage in the United States. The agency believed that as long as it was kept out of human food, Cry9C corn posed no health and safety problems for the environment or animals.

The EPA required AgrEvo to take specific steps to be certain that no StarLink corn would get into the human food chain. The agency required the company to enter into "grower agreements" with farmers who bought StarLink seed, telling them of the market restrictions on their harvested corn crop. Farmers were also required to plant a 660-foot "buffer zone" around their StarLink corn fields, to prevent StarLink corn pollen from cross-pollinating with normal food corn in nearby fields. Corn harvested from within the buffer zones was subject to the same marketing restrictions as StarLink corn. The EPA did not approve Cry9C corn for export.

AgrEvo, which was subsequently bought by the German-French corporation Aventis, hoped to convince the EPA at a later date to license StarLink corn for human food. Aventis officials realized that to change the EPA's mind about the safety of the product, however, they would need to convince the agency's scientific advisers that the Cry9C protein is not likely to cause allergic reactions in people.

In April 1999 Aventis submitted a petition to the EPA seeking approval for StarLink corn in human food. Company officials believed that the Cry9C protein was no more likely to cause allergies than any other Bt corn variety, and they hoped to convince the agency's scientific advisers to recommend EPA approve it for people to eat. The agency was being asked to predict whether the Cry9C protein would trigger allergic responses in sensitive people. The EPA seemed in no hurry to take up the question: it did not meet on the request for another ten months.

StarLink was the only variety of Bt corn that gave American officials pause, but many Europeans had serious questions about the safety of all kinds of Bt corn. While EPA officials grappled with cutting-edge scientific questions about the potential allergenicity of particular Bt plants, across the Potomac lawmakers on Capitol Hill turned their attention to troubling new trade statistics with the European Union.

SEVEN

SOUND SCIENCE,
STERILE SEEDS

I llinois representative Thomas Ewing had one burning question on his agenda when he called members of his agriculture subcommittee to order at ten-thirty on the morning of March 3, 1999: How dare the Europeans refuse to buy America's genetically engineered crops?

"The benefits of agricultural biotechnology are numerous," Chairman Ewing stated, opening the hearing in the Longworth Building, home to the powerful House Agriculture Committee. Michigan Republican Nick Smith agreed. "The biotech revolution underway is perhaps the most important single event in the history of agriculture. Not only will genetically modified products lead to decreased pressure upon the land, biotechnology also gives agricultural producers the technology to transform today's commodities into food and fiber products with new attributes and purposes." Genetically engineered plants, Smith said, will enhance human health, overcome environmental problems, improve the taste of foods, and lower grocery costs for consumers.

Department of Agriculture undersecretary August Schumacher was the first administration official called to testify. He predicted that the next great breakthrough in feeding humanity and saving lives around the globe would likely come from a scientist using biotechnology. "Just last week, researchers from Dartmouth in New Hampshire and the University of Newcastle in England announced a gene they have isolated that controls the iron content of plants. This would be an extraordinary first step in solving one of the world's most significant nutritional problems: iron deficiency, a problem so severe it is sometimes called hidden hunger," Schumacher told the committee.

"We have no reason to fear these products. We eat and drink them every day. We recognize that new technologies in any fields can impose risks. Yet, we must let sound science be our guide," the USDA official continued. "Objective, science-based regulations can provide the necessary protection that allow us to take advantage of the benefits that these new technologies offer."

By now, it had become clear to members of Congress that most European citizens were unwilling to eat the technologically "enhanced" foods being served up by American growers. For a host of reasons—from fears about health risks and ecological damage to religious qualms and moral objections—Europeans simply did not view crops bioengineered to ease weed management and pest control with the enthusiasm U.S. agricultural and biotechnology interests expected.

"It appears that our trading partners are rather arbitrary in restricting access to their markets for our biotechnology products," Ewing complained. "There does not seem to be a lot of good science out there to back that up—back up their failure to allow us into their markets with our products. We are drowning in corn. We ought to be getting it into other markets."

Recent trade figures painted a gloomy picture. In 1998 U.S. farmers lost about $200 million worth of corn sales to the European Union, as exports plummeted to 3 million bushels, down from nearly 70 million in 1997. This steep decline was a big problem for American corn growers like Roger Pine. A farmer from Lawrence, Kansas, Pine had come to Washington to testify in his role as president of the National Corn Growers Association.

"Mr. Chairman, we produce more than three million bushels of corn in Douglas County, Kansas," he began. "In Livingston County, Illinois, corn production is more than 37.5 million bushels, twelve times our total exports to the EU last year." Pine charged that Europe's refusal to approve new varieties of genetically engineered corn was politically motivated. "When science is subverted by politics and socioeconomic concerns, U.S. farmers don't have a chance."

Pine was unwavering in his support of biotechnology, however. "Plant genomics and biotechnology are critical to the long-term viability of U.S. agriculture," he insisted.

Michael Yost, a soybean and corn farmer from Murdock, Minnesota, was also high on biotechnology. Like his colleague from Kansas, Yost worried about the rejection of genetically modified foods by consumers in Europe and Japan. "In the three years since commercial introduction of genetically modified crops in 1996, agricultural biotechnology has become the defining issue in how the world will feed itself in the next century, and whether this decision will be based on scientific evidence," said Yost, president of the American Soybean Association. The very livelihoods of U.S. farmers were at stake, he told the subcommittee. "Also at stake is the viability of the basic rules of international trade, which are guided by science-based determinations. For U.S. soybean farmers, and for producers of other biotech crops, these stakes couldn't be higher."

Ewing's Subcommittee on Risk Management, Research, and Specialty Crops had only recently assumed jurisdiction over the issue of biotechnology in agriculture. An important job of congressional committees is to make sure the federal agencies charged with protecting the public are properly overseeing the nation's laws and regulations. Given the heated controversy in Europe—and to a lesser extent, among public interest groups in the United States—the congressmen might have profitably used the hearing to probe the safety tests that federal regulators required companies to perform on genetically engineered foods before approving them for sale. It was, in fact, the lack of safety testing by independent agencies that most troubled many Europeans about these new foods.

Representative Smith did raise the issue of safety testing briefly. The United States needs "to have good research and testing to make sure that these products are safe. What is out there? What are we doing?" Smith asked Gus Schumacher. "Should we be expanding our research and testing to prove, within a reasonable doubt, that these products are safe, rather than calling on the EU or any other country to prove to us that they are not safe?"

Schumacher apparently did not feel comfortable answering specific questions about the safety testing done on gene-altered foods, so he asked Richard Parry, a scientist with the USDA's Agricultural Research Service, to field the question.

"Proving the negative is very difficult. So far there have been no untoward effects that have been discovered," Parry answered. His response was perhaps less forceful than Smith might have liked.

Tim Galvin, an administrator with the USDA's Foreign Agricultural Service, volunteered that 70 percent of the plant scientists in the United States worked for private companies. This made it hard for the government to find qualified experts to do objective, independent safety reviews of genetically engineered crops.

"Some have raised concerns over the safety of genetically modified products. USDA has a very effective program which regulates crop and livestock biotech products to ensure that no 'rogue' products are released to the environment. EPA and FDA also regulate biotech products. Combined, these regulatory controls are the most stringent in the world," Smith asserted. "No other country requires such comprehensive tests and proof of the absence of problems before a product is made available to the general public."

Much of the hearing was devoted to berating European countries for failing to push these new biotech foods through their own regulatory bodies as quickly as U.S. agencies had done. While a company could secure final U.S. regulatory approval for a biotech product in an average of nine months, EU approval time averaged between eighteen months and two years, according to USDA officials. After March 1998 the European Union had essentially stopped approving new genetically modified food varieties. What the European Union needed to break the logjam, the congressmen and the Clinton administration witnesses agreed, was an FDA of its own, modeled after the U.S. agency.

Representative Gary Condit, a Democrat from California, acknowledged that ultimately the success or failure of genetically modified crops would be decided in the marketplace. "Past experience indicates that consumers can effectively sort through misinformation, and decide about the real value of new goods and services. When color TVs were introduced, there were warnings about potential risks from mutations," he noted. "With microwave ovens there were warnings about risks of abnormalities from radiation. The widespread adoption of both technologies suggests that when scare tactics lack solid scientific evidence, they tend to be short-lived. This will, no doubt, prove true for agricultural biotechnology as well," he predicted confidently.

Still, American growers were losing commodity sales to Europe now, and U.S. trade representatives were picking up foreboding warnings that consumers in several other countries, including Japan and Korea, might follow the European lead. The congressmen wanted reassurances that the Clinton administration would do something to make these errant nations accept genetically engineered food exports from the United States.

Representative Condit asked Schumacher what the United States would do if the European Union were to ban genetically modified imports. The undersecretary responded that the administration would "reserve all of our trade negotiating rights through the WTO."

Jim Murphy, an official with the Office of the U.S. Trade Representative, elaborated, saying that "you would obviously have to know why they put the ban in place. It is unimaginable to us that there would be a good science-based reason for doing that—in which case, they would be vulnerable under the World Trade Organization."

The strategy the administration had settled on—with the support of Congress, the food processing industry, and agribusiness—was to enforce the application of a single concept: "sound science." Under rules written by the World Trade Organization (WTO), a nation cannot refuse to import a product that has passed the scientific safety reviews of various recognized institutional bodies. A nation cannot refuse to import foods solely on the basis of ethical beliefs, for instance. To do so is to risk a lawsuit by the exporting nation, on the grounds that the importing nation has raised an unfair, nontariff barrier to trade.

In the spring of 1999 Europe and the United States were already engaged in a trade dispute over hormone-treated beef. The European Union refused to import beef that had been treated with hormones, claiming that the hormones posed a potential threat to public health. The United States refuted the claim, arguing before a WTO dispute-settlement body that there was no scientific evidence that the hormones used by U.S. cattlemen were unsafe. The USDA and the FDA had approved the safety of three natural hormones and three synthetic hormones for use as growth promoters and fattening agents in cattle.

The WTO ruled in favor of the United States, but the European Union showed no signs of coming into compliance with the ruling.

The phrase "sound science" is chanted like a mantra in venues throughout Washington: at EPA meetings, at seminars sponsored by conservative think tanks, at food industry conferences, and during other congressional hearings. The concept is invoked most often by industry trade groups or their boosters in Congress when risky products or environmentally destructive practices come under attack by public interest groups. One of the biggest battles of the 1990s fought over the issue of sound science centered on the cause of global warming. An overwhelming majority of scientists from industrialized nations compiled data throughout the decade showing that the Earth is undergoing climate change because of fossil fuel emissions. Yet several conservative think tanks charged that "sound science" did not support this claim. The oil and coal industries pointed to the work of a handful of scientists—often supported by energy trade groups—who drafted papers and delivered expert testimony that purportedly refuted both the well-documented warming of the atmosphere and its likely cause.

Like industry groups, most environmental and consumer groups also have scientists on staff and consult with university researchers to generate facts and figures in support of their positions. Pro-industry advocates of "sound science" often dismiss the scientific data generated by their opponents, calling it "junk science."

Proponents of genetically engineered crops—and the big multinational corporations that sell the seeds—argue that since scientists have turned up no evidence yet that anyone has been harmed, and no evidence that the environment has been endangered, the new foods must be safe. Once the premise of safety has been accepted, many corollaries follow. Because the gene-altered foods are safe, opponents of the technology have no right to limit their use in any fashion. Because U.S. regulators deem these foods to be "substantially equivalent" to their traditional counterparts, opponents lack ample justification to seek mandatory labels. Without labeling and segregation, genes from bioengineered crops will gradually flow into traditional varieties, contaminating seeds. Over time, inevitable mishaps in seed handling and cross-pollination of traditional plants by gene-altered varieties growing nearby in open fields will lead to the blurring of distinctions between the two kinds of foods—the outcome sought by biotech companies.

Citizens and scientists who oppose the genetic engineers' view of "sound science" counter that when it comes to potential adverse effects from newly introduced technologies, "absence of evidence is never evidence of absence." Environmentalists recall a list of earlier supposedly harmless technologies that turned out to have lethal side effects—long after they had been allowed to proliferate.

The development of nuclear energy provides a clear example of the limitations of science and scientists to predict outcomes of a new technology. In the 1970s, with some one hundred nuclear reactors built, commissioned, or under construction in the United States, the public became increasingly worried that a nuclear power plant would suffer a catastrophic accident and release radiation. To quell public fears, the Nuclear Regulatory Commission contracted a team of scientists to use an analytic technique known as fault tree analysis to identify all the errors and malfunctions that could cause a nuclear power plant accident. The scientists then calculated the likelihood of each sequence of events. The resulting reactor safety study, known as the Rasmussen Report, concluded that the probability of a severe accident was one in a billion per reactor per year of operation. Public skepticism about these "scientific" calculations was borne out in 1979, when an accident at the Three Mile Island reactor in Pennsylvania nearly resulted in a meltdown. Though no one was killed, the credibility of the nuclear industry plummeted. Seven years later, in 1986, the Chernobyl power plant exploded, releasing into the atmosphere hundreds of times more radiation than the combined atomic bombs the United States dropped on Hiroshima and Nagasaki.

Many ecologists and environmentalists in Europe—and in the United States—also questioned the soundness of a narrowly defined science that fails to take uncertainty into account. Ecologists, whose fieldwork takes them out into nature, where soil bacteria, plants, and animals interact, often have a very different perspective on genetic engineering than molecular biologists, who work in sterile laboratory conditions, with test kits, enzymes, and genes.

While uncertainties about the safety of genetically engineered foods for humans and the environment lay at the heart of Europe's rejection of them, representatives at the 1999 congressional hearing explored other possible explanations as well. The congressmen formulated two general theories to explain Europe's aversion. Representative Earl Pomeroy of North Dakota suggested that Europeans have different "cultural" attitudes from Americans. He did not elaborate.

Certainly many cultural differences exist between Europe and the United States. Food is much more than simply the calories and nutrients that sustain our bodies. How we eat, what we eat, and even when we eat all go into making up who we are, individually and as a people. Each European country draws part of its national identity from its cuisine and eating habits. Traditionally, the Spanish have a siesta after lunch, begin snacking on *tapas* when the average American is sitting down to dinner, and then dine at ten o'clock at night. The French have set mealtimes for lunch and dinner, which are leisurely social affairs, and they seldom snack in between, unlike Americans, who tend to grab food on the run or sit down for a meal at whatever hour of the day fits their work schedules. Still, some generalizations can be made about Europe as a whole. Europeans view food differently from Americans—more reverently, perhaps; Americans, it is often said, will eat just about anything. Many Europeans, even city dwellers, have close ties to friends and relatives in the countryside; many grow kitchen gardens. Throughout Europe patches of farmland are never distant from cities and towns, whereas many Americans are far removed from the vast swaths of farming regions of the country where most of the nation's food is grown. Europeans also spend, on average, a larger portion of their incomes on food than Americans do.

Chairman Ewing seemed convinced that European leaders were using "cultural" objections merely as a smokescreen. Their real motivation in rejecting genetically modified foods, he suggested, was protectionism and a desire to keep American crops out of Europe. This argument, however, did not explain why Europeans had rejected Roundup Ready soybeans. Very few soybeans are grown in Europe, and buyers there had to scurry to find sources abroad willing to supply them with nonengineered soybeans. Some European buyers had turned to Brazilian suppliers, but others were contracting with American farmers who were willing to guarantee delivery of nonengineered soybeans.

Representative Michael Simpson of Idaho asked Trade Representative Murphy if Europeans have "a cultural problem" with genetically engineered foods. "When you talk about genetically altered, genetically engineered products, it conjures up all sorts of weird things in your mind," he said. "Do you think that is a real problem in their culture?"

Murphy responded, candidly:

I think culturally there is just inherently a more conservative attitude there. I joke with some of my European friends saying that the definition of an American is a risk-taking European. We immigrated here. We had faith in the future. Those who stayed behind had a little less faith in the future. They were more risk-averse. You see this in endless conversations with Europeans that often comes down to them saying, you cannot tell me or prove to me that somewhere, long term, we will not find something wrong with this technology. Of course, you cannot. You cannot prove the negative. Our response, of course, is you have to go with the best science available, and nobody is finding, in the current research, any problems with this technology. It is completely safe. There are 270 or 280 million Americans eating this stuff every day. We are all walking around very healthy. So, you Europeans have a laboratory here.

Many Europeans have found little reassurance in the food experiment unfolding in the United States. According to Alan Simpson, a Labor member of Parliament in the UK, "From the European perspective, it's almost as if we've looked in awe at what's happening in the U.S., either as an act of madness or with huge admiration that the society is willing to offer its entire population as a human laboratory."

On February 12, 1999, two weeks before Thomas Ewing's agriculture subcommittee hearing, twenty scientists from thirteen countries had released a memorandum supporting Arpad Pusztai's work. Pusztai's potatoes again exploded onto the front pages of British newspapers, this time with renewed ferocity. "This scientist revealed the perils of GM food. Now he has been gagged for life," screamed one headline. "A doctor destroyed for being in the right," blared another. The British media published a rash of stories charging that political corruption and multinational corporate influence were to blame for the shabby treatment accorded to Arpad Pusztai.

In the week from February 12 to 18, 1999, genetically modified food was Britain's top-ranked news story, having the most combined total column inches in the country's leading newspapers—*The Guardian, Times, Daily Telegraph, Independent, Sunday Times, Observer,* and *Daily Mail*. Stories about genetically modified foods topped news about President Clinton's acquittal in the U.S. Senate impeachment trial and the bombing campaign in Kosovo, one biotech industry official found.

Public opinion in Great Britain, which was already highly suspect of biotech foods, hardened. On February 16, 1999, the gag order that government officials had imposed on Pusztai shortly after his controversial television interview in August 1998 was lifted, allowing the scientist to tell Parliament and the general public about the problems he found with rats that had been fed experimental genetically modified potatoes. Two days later, on February 18, Greenpeace activists dumped four tons of gene-altered soybeans outside 10 Downing Street, the official residence of Prime Minister Tony Blair. "We are taking these GM soya beans to one of the few homes in the UK where they want to eat it," a spokesperson for the environmental group said.

Blair published an article in *The Daily Telegraph* on February 20, 1999, saying the government "stands firm" on its position in support of genetically modified foods. "We need to be governed by good science not scare mongering," he wrote. "I believe, in the end, that reason will prevail." The Royal Society and other scientists also stepped into the fray, trying to reassure the public that genetically engineered food was safe to eat. But advocates of gene-altered foods found themselves greatly outnumbered by opponents. A poll conducted by *The Independent* on Sunday, February 21, found that 68 percent of Britons were worried about eating genetically modified foods, while 59 percent said they were unhappy with the way the government was handling the issue.

In late February 1999 the Local Government Association recommended a ban on genetically modified foods in Britain's schools, hospitals, town halls, and old people's homes. On March 3 the director of the National Pollen Research Unit at University College in Worcester said the risk that pollen from transgenic plants would contaminate conventional crops may have been seriously underestimated. In mid-March Food Minister Jeff Rooker announced new regulations requiring cafés and restaurants to warn customers when food contained genetically modified corn or soybean products. The UK food industry's support for biotech foods began to crumble, as one retailer after another pledged to remove those ingredients from their home-brand products.

In Ireland, meanwhile, a high-profile trial opened in New Ross District Court on March 30, 1999, pitting seven Irish citizens against Monsanto. In the first case of its kind in Irish history, the citizens were charged with trespass and criminal damage to a test plot of Monsanto's genetically engineered sugar beets. The seven people charged had damaged experimental plants growing on a quarter-acre plot on a farm near Arthurstown in County Wexford on June 21, 1998.

The protesters included Green Party members and John Seymour, a prominent author and organic farmer. Seymour, eighty-four years old, said on the Voice of Irish Concern for the Environment website that "when a huge multinational corporation comes and starts planting completely untested and untried genetically mutilated plants in the country where I have made my home, and the government agencies which are supposed to be there to protect us from that sort of thing fail in their duty, then I feel it not only my right but my duty, too, to do something to try to stop them. And if I have to go to prison because of it, then I will go with a good will, and make the best of it, and when I get out I will try to stop them again!"

Following a two-day hearing, the judge found that the charges against the Arthurstown Seven were proven. But he reduced the amount of damage payments from the £16,000 sought by Monsanto to just £30. The judge said he believed that the protest had been conducted "in a very honest, good-humored manner," and the beliefs of the protesters were "honestly held," *The Irish Times* reported.

Finally, in May 1999, in a decisive blow to genetically engineered foods, the British Medical Association (BMA) released a paper disputing the U.S. FDA's justification for presuming that those foods were safe either for human consumption or for the environment. "The BMA believes that more research on issues around allergenicity and possible toxicity of GM foodstuffs needs to be undertaken," the fourteen-page policy paper stated. "Information about the effect of genetic modification on the chemical composition of food and in particular its safety is needed urgently."

The BMA, which represents more than 115,000 British physicians, said that "we cannot at present know whether there are any serious risks to the environment or to human health involved in producing GM crops or consuming GM food products. Adverse effects are likely to be irreversible; once GMOs are released into the environment they cannot be subject to control."

The British doctors questioned the scientific basis for the U.S. FDA's doctrine of substantial equivalence, stating, "This concept does not account for gene interaction of unexpected kinds which may take place in GM foods. The possibility that certain novel genes inserted into food may cause problems to humans is a real possibility, and 'substantial equivalence' is a rule which can be used to evade this biological fact."

Throughout the spring of 1999, the U.S. government seemed to be in denial about the intensity of European opposition to genetically modified foods. The Clinton administration stepped up diplomatic efforts to force the European Union to approve new crop varieties. U.S. farmers set about planting biotech seeds on a staggering 75 million acres—about one-quarter of the nation's cropland.

On March 15, 1999, Undersecretary of State for Economic, Business and Agricultural Affairs Stuart Eizenstat predicted a rosy future for gene-altered crops. "Within a few years, virtually 100 percent of U.S. agricultural commodity exports will be genetically modified or mixed with GMO products," he told the Senate Finance Committee.

When the Biotechnology Industry Organization (BIO) opened its annual meeting in Seattle on May 17, 1999, agricultural biotechnology leaders appeared unfazed by the overwhelmingly negative response to genetically modified crops throughout Europe. Scientists and industry executives spoke exuberantly about the coming "second wave" of gene-altered foods, confident that these valuable new products would bring the Europeans around.

"We're looking forward very soon to tomatoes with enhanced phyto-nutrients, called lycopenes, to prevent prostate cancer, cooking oils modified to contain vitamin A to treat night blindness, a condition that affects many people in developing countries, and bananas and potatoes that contain vaccines against childhood diseases," Carl Feldbaum, BIO's president, told the 5,500 delegates gathered at the airy Washington State Convention and Trade Center. BIO represents more than nine hundred companies engaged in all aspects of biotechnology, from agriculture to pharmaceuticals and other medical applications.

Roger Beachy organized a session on the "fantastic visions of biotechnology's possibilities" that emerge when scientists imagine. Products already in the biotech pipeline, he said, included edible vaccines, naturally colored cotton, and low-fat French fries. Beachy, one of the deans of genetically engineered foods, in the mid-1980s was a leader of the joint Washington University and Monsanto research team that engineered a virus-resistant tomato. On January 1, 1999, he had been named president of the new Donald Danforth Plant Science Center in St. Louis, a research center funded by the Danforth Foundation, Monsanto, and the state of Missouri.

William White, who heads up a division of Monsanto called Integrated Protein Technologies, described the progress the company was making in transforming green plants into living factories that would be able to cheaply and efficiently manufacture medicines. Corn, soy, and tobacco were the company's stainless-steel plants, White said, noting that plants had the best potential to be the source for the next generation of pharmaceutical products.

A big selling point for using green plants to produce complex proteins—instead of concrete and metal production facilities—is the low capital investment needed to start production of a new product. Another benefit of using plants, rather than human or animal sources, for the production of therapeutic proteins is that they do not carry the same risk of harboring pathogens and diseases, White noted.

Steve Benoit of Kellogg Company talked about a new line of "functional foods" under development, which he defined as "great food that does great things." A new Kellogg subsidiary called Ensemble

Functional Foods Company was in the midst of researching ways to use biotechnology to develop foods and beverages that would provide a physiological benefit beyond basic nutrition, to manage or even improve health. Two factors were driving the desire for such products in the marketplace, he said: the aging population and changes in medical care. Doctors were spending less and less time with their patients, so consumers were looking for ways to improve their health on their own. Ensemble Functional Foods planned to focus on foods designed to manage health problems such as high cholesterol, heart disease, diabetes mellitus, and colorectal conditions. "Healing the world with food is our vision," Benoit added.

Functional foods are an important part of the second wave of genetically engineered foods that the biotech industry is promising. Researchers are keen to identify the many substances in food plants that are thought to retard the growth of cancer cells, for instance, and then manipulate the plants to significantly boost production of those substances.

Channapatna Prakash, who directs the Center for Plant Biotechnology Research at Tuskegee University in Tuskegee, Alabama, described his efforts to engineer sweet potatoes to express proteins. Sweet potato, which Prakash said is the seventh most important crop in the world, is popular among poor farmers because of its ability to thrive in poor soils and tolerate drought. It has a high yield and is a major component of the diet of people in Uganda, New Guinea, and many other developing countries. But Prakash characterized the protein quality in this staple crop as poor. Most plants, in fact, do not contain the essential amino acids found in meat, making them less adequate sources of protein than animal sources.

Prakash and his research team had introduced a synthetic protein gene into sweet potatoes that "was designed from the ground up to be stable in plants," he told the BIO meeting. The protein made up more than 12 percent of the new transgenic sweet potato, compared with only 2 to 3 percent of ordinary sweet potato plants, he said.

The convivial atmosphere at the BIO meeting was interrupted briefly when a few dozen antibiotechnology protesters marched into the convention center. Some wore costumes depicting mutant fruits and vegetables. Others carried signs reading "Look Out for Frankenfood," "Buy Organic," and "Operation Cremate Monsanto." Conference-goers attired in conservative business suits looked on uneasily as the demonstrators moved slowly up four flights of escalators, beating drums and chanting, "Save food now. End the corporate plow," and "Our genes are not for sale."

The demonstrators exited the building and staged a rally on a glassed-in veranda, within full view of the BIO delegates. I followed them outside. The people I interviewed complained that U.S. newspapers and television stations were completely ignoring American opposition to genetically engineered foods.

"We had a huge Earth Day demonstration in San Francisco and Berkeley at four different supermarkets, and not one TV station covered it," a young woman wearing a "Super Hero" outfit and identifying herself only as Biogirl told me. "Most of the shoppers we talked to had never heard of genetically engineered food. They were shocked when we told them about it," she said. "We got a lot of support and encouragement from people."

"We're here to show our opposition to agro-biotechnology," said another protester, whose group had been interviewed by a German television station. "We want the U.S. media to know it isn't just Europeans who oppose genetically engineered food. We're against it too."

In a church a few blocks from the convention center, the protesters were holding a counterconference to the BIO meeting, provocatively dubbed Biodevastation3. The following day I attended the gathering of about 150 environmentalists, health food store owners, political organizers, ecologists, and students hunkered down over organic bean-wrap sandwiches in the Plymouth Congregational Church. Some had come to the teach-in to learn more about the science of genetic engineering. Others had already decided it was a dangerous technology that, if allowed to proliferate on the planet, could spell a literal end to the nature of people, plants and animals as we know them.

Brian Tokar, a faculty member at the Institute for Social Ecology in Plainfield, Vermont, who holds a master's degree in biophysics from Harvard University, energized the participants, reminding them that U.S. activists had been in the forefront of protests against genetically modified bovine growth hormone when Monsanto first introduced it. At the end of the decade, though, European activists had captured the lead in the fight against genetically engineered food. "We have a little catching up to do," he exhorted the audience.

"We're seeing new scientific discoveries being rushed to the market-place at a pace never seen before," Tokar charged. The multinational corporations that developed the gene-altered corn, soybeans, and other products in our kitchens, he continued, started selling them pre-maturely, before they were adequately tested, in order to recoup their substantial research and development costs and start making a profit. "When people come to understand the implications of genetic engi-neering, they overwhelmingly reject it."

Martha Crouch, a professor of biology at Indiana University for twenty years, described herself as an ex–genetic engineer. During her first decade at the university, she had been a preeminent molecular biologist who conducted research on the development of seeds and pollen in canola, soybeans, evening primrose, and corn. But in 1990 she developed misgivings about the entire enterprise of reengineering life and turned her back on her research. Instead she began to study and teach courses on traditional agricultural systems. Crouch said she also served as a wild mushroom inspector at the Bloomington Farmer's Market.

Crouch told the counterconference that the discipline of genetic en-gineering was built on the idea that "the human has a plan for an or-ganism that previously had a life of its own, a plan of its own within its own species." Standing at the podium, she held up a large white mushroom and described the subtle features about its cap, its stem, its gills, and the way the stem joins the cap that distinguish this edible mushroom from a poisonous variety that might fool the eye of an un-trained observer. Through knowledge and experience, Crouch said, she was able to observe the characteristics that made the one mush-room safe to eat.

Genetic engineers, Crouch charged, had made it impossible for even a careful observer to distinguish a genetically altered food from its natural counterpart. Because only a sophisticated laboratory DNA analysis could differentiate a genetically engineered Bt potato from a normal Russet Burbank potato, for instance, vital knowledge had been taken away from the farmer and the consumer. "I can take the risk, knowingly, with eating wild mushrooms, but with potatoes, genetic engineers have taken my knowledge away for their own greed," she said. "I can't tell the genetically engineered potatoes from regular potatoes."

The scientists and environmental activists addressing Biodevesta-tion3 voiced skepticism about the hypothetical benefits of the second wave of products being promoted a few blocks away at the BIO conference. Many were more worried about problems that might have gone unnoticed, or unpublicized, in the gene-altered foods already on our dinner plates. Others were interested in how new patented crops would affect poor farmers around the world.

Edward Hammond, a researcher at the Rural Advancement Foundation International (RAFI) in North Carolina, described one of the most controversial creations of the agricultural biotechnology revolution: the "Terminator." On March 3, 1998, Delta and Pine Land Company, an American seed company, and the U.S. Department of Agriculture were jointly awarded U.S. Patent Number 5,723,765, titled "Control of Plant Gene Expression." The patent, which Hope Shand, RAFI's director of research, cleverly dubbed the "Terminator," is a technique for making the seeds of genetically engineered plants sterile. Assuming the patent can be made to work as described, crops engineered with terminator genes would kill their own seeds in the second generation, thus making it impossible for farmers to save and plant their harvested seeds. RAFI is a small, Canada-based organization that studies agricultural genetic diversity and the effect of biotechnology and patents on rural societies.

In May 1998 Monsanto entered into an agreement with Mississippi-based Delta and Pine Land to acquire the cotton seed company for $1.8 billion. The acquisition would have given Monsanto access to the powerful new technology, which biotech companies refer to as a "technology protection system." Hammond warned that Terminator seeds could hurt the estimated 1.4 billion farmers in the developing world who still practice agriculture in the time-honored way: saving a portion of their harvested seeds for the following planting season. If Monsanto were to engineer the Terminator genes into rice and wheat—staple foods for the majority of the world's poor—subsistence farmers, RAFI and other activists charged, would become beholden to the large international biotech seed companies that control the technology.

"The key to Terminator is the ability to make a lot of toxin that will kill cells, and to confine that toxin to seeds," Crouch wrote in a paper entitled "How the Terminator Terminates," published by the Edmonds Institute in Edmonds, Washington. To program a plant to accomplish this totally unnatural feat, timing is key. The engineered plant must grow normally and produce seeds that mature normally, right up until the last possible moment. Otherwise the crops would fail to produce a marketable yield and would be of no value to the farmers planting them. In a cotton plant, Crouch explained, genetic engineers would take the promoter from a gene that is normally activated late in seed development and fuse it to a gene that codes for a protein that would kill an embryo going through its very last stages of development.

Hammond said RAFI had also discovered several recent patents filed by major biotech companies describing a companion plant engineering technique. Officially called "genetic use restriction technology" by biotech firms, the group called this technique the "Traitor." It involves altering the genome of a plant so that it depends on a particular chemical agent to "turn on" or "turn off" genetic traits, Hammond said. Farmers who were to plant seeds containing Traitor genes would have to purchase proprietary fertilizers or other chemicals to make their plants express the desired trait.

Companies could also use the Traitor technology to disable a plant's natural functions, including the ability to fight diseases, Hope Shand wrote in an article posted on RAFI's website. "The long-term implications, both for farmers and for national seed sovereignty, are sobering. National agricultural production could become wholly dependent upon foreign exports of critical chemical inducers."

Had RAFI not uncovered the Terminator patent in July 1998 and publicized it widely, multinational biotech companies might have succeeded in quietly disabling the ability of the world's major crops to reproduce naturally. Taking advantage of the power of the Internet, RAFI disseminated information about the Terminator and Traitor technologies to governments and to environmental, agriculture, and food security organizations around the world. The small group sent letters to the agriculture and environment ministers in 140 countries, entreating them to ban seed sterilization technologies.

Monsanto's bid to buy Delta and Pine Land catapulted the company into the storm swirling around the controversial Terminator technology. Already vilified in Europe for slipping genetically engineered soybeans into the food chain unlabeled, the global giant came under fierce attack in India, where hundreds of millions of small family farmers rely on the age-old practice of planting saved seeds. Hundreds of Indian farmers were so enraged by press reports of the Terminator gene that they set fire to fields where Monsanto was growing experimental Bt cotton crops. These plants did not contain the controversial technology, but the Terminator gene had become so firmly linked with the company in the popular imagination that a group of farmers torched a Bt cotton test site in Karnataka on November 28, 1998. A few days later farmers burned a second Bt cotton field in Andhra Pradesh.

The revelation that private biotechnology companies were using the tools of genetic engineering to rob seeds of their innate ability to reproduce unnerved people around the globe. Seeds, by definition, are the source from which new life springs, nature's vehicle for perpetuating plant life. In 1999 Indian farmers led a caravan through Europe, demonstrating against Monsanto's genetically engineered crops at WTO headquarters in Geneva, the European Commission headquarters in Brussels, and the Organization for Economic Cooperation and Development in Paris. Wealthy Europeans found common ground with impoverished Indian farmers in the international fight to halt the spread of Monsanto's crops and the Terminator technology.

Back at the BIO meeting, the biotech company executives seemed strangely insulated from the fierce ideological war their food products had provoked around the globe. Food industry polls continued to show that few Americans had any idea they were eating genetically engineered foods. The industry had no reason to believe that the "hysteria" gripping Europe would hop the Atlantic and infect U.S. consumers.

But on May 20, 1999, the final day of both the BIO and the Biodevestation3 conferences, the American public awoke to news of a frightening study. When I picked up *The New York Times* at a newsstand in Seattle on Friday morning, a vivid photo of an orange-and-black barred monarch butterfly perched on a yellow flower leaped off the front page. Above the stunning picture, the headline read: "Altered Corn May Imperil Butterfly, Researchers Say."

The article described a simple laboratory experiment showing that pollen from Bt corn kills monarch butterflies. A Cornell University entomologist found that nearly half of the monarch caterpillars fed on a diet that included genetically engineered corn pollen died within four days, while none of the caterpillars fed pollen from conventional, non-engineered corn died.

At the convention center, as BIO exhibitors broke down their displays and delegates toted their luggage to the final sessions before flying home, the mood seemed subdued. Three blocks away, though, the atmosphere at the congregational church crackled with buzz about the monarch butterfly study. Environmentalists at the Biodevastation3 meeting warned that the dark side of genetic engineering was emerging. The EPA had assured Americans in 1995 that Bt corn would kill only European corn borers and would not harm wildlife. Ronnie Cummins, director of the Organic Consumers Association, brandished a copy of *The New York Times* in the air, shouting, "The biotech century? It isn't going to happen."

The genetically engineered food war was about to come home.

EIGHT

LETHAL CORN POLLEN

On a hot afternoon in late July 1998, John Losey was scouring a thicket of weeds in Freeville, New York, for bugs. The young entomology professor was looking for one species of insect in particular: the European corn borer, which might be feeding on the leaves of wild plants surrounding cornfields. One of the most common plants he came upon was milkweed. "There's a lot. It's a very conspicuous weed. And I found that when the corn sheds pollen, the milkweed leaves, like most other plants that are around cornfields at that time, get dusted with that pollen," he told me during a visit to his laboratory on the Cornell University campus. This observation started him wondering: "What would happen if that were Bt pollen being dusted on those leaves?"

Losey had never worked with monarch butterflies, but like most biologists he knew that monarch caterpillars feed exclusively on leaves of the milkweed plant. Most creatures shun milkweeds because they produce a thick white sap that contains noxious substances called cardiac glycosides, or cardenolides. Birds and other animals become sick when they consume even a small quantity of cardenolides. Over the course of plant and insect evolution, however, monarch caterpillars developed a unique capacity to eat milkweed leaves without becoming sick. The adult monarch butterfly feeds on the nectar of flowers, but the cardenolides accumulated during its earlier stage of life remain in its tissues. The bright orange and black coloration of the monarch serves as an unmistakable warning sign to predators; once sickened by a monarch caterpillar or butterfly, most birds and other animals avoid them. Other species of butterflies that took on the orange and black markings, mimicking monarchs, enjoyed protection as well.

John Losey decided to find out how monarch caterpillars would fare on a diet of milkweed leaves coated with Bt corn pollen. In August 1998, with the help of researchers in Cornell University's plant breeding division, he collected pollen from both genetically engineered Bt corn and from conventional, non-Bt corn plants. His colleague, Linda Rayor, provided him with monarch caterpillars. The two scientists set up a simple feeding study in a growth chamber on the fourth floor of Comstock Hall, which houses Cornell's entomology department faculty. There they fed milkweed leaves to three groups of three-day-old monarch caterpillars. One group of larvae ate milkweed leaves dusted with Bt corn pollen, a second group ate leaves dusted with regular corn pollen, and a third group ate plain milkweed leaves with no pollen.

Two days into the feeding experiment, the caterpillars eating the leaves with Bt corn pollen started to die. "It was about eighteen percent mortality—and none in the control or the regular pollen treatment," Losey said. At the end of four days, 44 percent of the monarch caterpillars that had been fed a diet of leaves dusted with Bt corn pollen were dead. None of the caterpillars fed either plain milkweed leaves or leaves dusted with regular corn pollen died.

When I asked Losey if he was surprised, he answered, "I just didn't know how this was going to turn out. I don't think I would have been terribly surprised either way because there's a lot of variation in Lepidoptera—in how susceptible they are to the Bt toxin itself." The order Lepidoptera numbers about 200,000 species: 20,000 butterflies and 180,000 moths.

"We really didn't know whether they would consume the pollen and be able to break that down enough so they would be harmed," Losey explained. "There were two things that could vary a lot. One is, if they get the toxin in their gut, are they going to be susceptible to it? And before that can happen even, would they consume the pollen? Would they digest it enough that they would get the toxin within their guts or are they going to just pass it through?"

The researchers ended the experiment after four days because they were running out of *Asclepias curassavica* milkweed leaves, but Losey assumes that more caterpillars would have died. "The larvae that were on the Bt pollen treatment were quite a bit smaller than the others, so our assumption is that death would have continued."

Losey was aware of the profound implications of his study. When the EPA granted approval for farmers to plant Bt corn in 1996, agency scientists stated that the toxin produced by the gene-altered plants would not harm honeybees and ladybugs and other beneficial or so-called "nontarget" insects such as butterflies. The Bt corn was engineered to kill the larvae of one specific insect pest, the European corn borer. In 1998, when Losey performed his experiment, an astounding twenty million acres, accounting for about 25 percent of the U.S. corn crop, were planted with Bt corn. Across the farm belt, as in the Finger Lakes region of New York, milkweed commonly grows around the edges of cornfields, within what scientists call the "corn pollen shadow"—the area where wind-borne pollen is carried and deposited on plants. Nearly half of the North American summer monarch butterfly population is concentrated in the midwestern corn belt, placing substantial numbers of monarchs at potential risk for ingesting toxic Bt corn pollen.

John Losey shared the results of his experiment with colleagues at a meeting of specialists in Bt corn and corn borers in January 1999. He then wrote up his findings and sent the paper to thirteen "friendly" reviewers for comment.

"You generally don't send it to nearly that many people. Usually, you send it to one or two," he told me. "But since this clearly was going to be a major issue, we wanted as many comments as we could to make sure we were covering all the bases."

Losey first submitted the paper to *Science*. The American journal *Science* and the British journal *Nature* are two highly regarded weekly science magazines that publish research results of interest to a broad range of scientists as well as the general public. Both journals have editorial boards that screen submissions before sending them out for peer review by scientists. Only newsworthy pieces or research that might appeal to a wide audience are peer reviewed.

"I had published in *Nature* before, and I wanted to publish in *Sci-*

ence," Losey said, "so I sent it to *Science,* and they came back and said, no, this is not exciting enough. They never sent it out for peer review."

Next, he sent the paper to *Nature.* The editors there decided the research was of broad interest and sent the paper, "Transgenic Pollen Harms Monarch Larvae," to referees for review. On March 23, 1999, *Nature* faxed Losey an acceptance letter—along with comments from two anonymous reviewers. One of the referees began his review: "Quite a startling finding with important policy implications."

When the eight-paragraph paper appeared in the "scientific correspondence" section of *Nature* on May 20, 1999, it ignited a media firestorm. News of the unintended harmful effect of Bt corn pollen raced around the globe several hours before the magazine came out, spread by wire services and the Internet. Dozens of U.S. daily newspapers ran prominent stories on Losey's study. "Bioengineered Corn Pollen Can Kill Butterflies," heralded *The Baltimore Sun;* "Genetically Altered Corn Shown Fatal to Monarch Butterflies," announced *The Buffalo News.* "Altered Pollen Kills Butterflies," cried *The Denver Post;* "Man-made Corn May Do More Harm Than Good," *The Orlando Sentinel;* "Gene-spliced Corn Imperils Butterfly," *The San Francisco Chronicle.* Alarming reports about a new kind of corn that makes pollen lethal to monarch butterflies was the first news many Americans had read about gene-altered food.

For consumers, environmentalists, and government officials from Toronto to Tokyo, this startling news suggested there were serious gaps in U.S. safety testing of gene-altered crops. One Canadian environment official reportedly called the finding "scary." He had assumed that the pollen from gene-altered corn did not contain the active Bt toxin. European Union officials halted approvals of all pending applications for genetically engineered seed corn. In Japan, Korea, Thailand, and Australia, the public began to ask to see scientific evidence proving the long-term health and environmental safety of biotech crops. After learning that the U.S. government had not required such studies, consumers throughout the Asia-Pacific region joined Europeans in demanding mandatory labels on foods containing genetically modified ingredients.

"It's time to stop looking at agricultural biotechnology through rose-colored glasses. This is a novel technology that poses serious risks to the environment," Margaret Mellon, director of the Union of Concerned Scientists' agriculture and biotechnology program in Washington, D.C., said on May 20. She added that the EPA had rushed into allowing twenty million acres of this corn to be planted without thoroughly assessing its risks. Mellon, who holds a Ph.D. in molecular biology and a law degree, had been working to slow the spread of biotech crops since 1987, when she joined the National Wildlife Federation in Washington, D.C.

In 1996, when the EPA approved widespread planting of Bt crops, Mellon and her colleague Jane Rissler had been among the first scientists to warn that plants engineered to manufacture pesticides would likely harm other organisms besides the intended pest targets. Rissler, who earned her doctorate in plant pathology at Cornell University, spent four years at the EPA, where she helped to develop biotechnology policies. In 1993, she and Mellon joined the Union of Concerned Scientists to set up an agriculture and biotechnology program. The two women had relentlessly dogged EPA and USDA bureaucrats, submitting dozens of scientific critiques of the biotechnology industry's plans for gene-altered corn, soybeans, potatoes, squash, and other foods.

"The fall migration of monarch butterflies from Canada to central Mexican forests is a natural wonder," Rissler said in a press statement issued in response to John Losey's research. "The loss of monarchs is too high a price to pay for engineered corn."

The finding that Bt corn pollen can kill monarchs led Rissler to wonder whether other vulnerable lepidopteran species might also be at risk from consuming the toxic pollen. She asked EPA and Fish and Wildlife Service officials to consider the potential threat to some of the nineteen species of butterflies and moths on the U.S. Endangered Species List that might be exposed to toxic Bt pollen while feeding on wild plants near cornfields.

The Environmental Defense Fund (EDF) in New York identified several rare species of butterflies that inhabit states where they might suffer mortality from Bt corn pollen. Rebecca Goldburg, a senior scientist with the EDF, a national environmental group with 300,000 members, charged that the EPA had failed to consider the risks of toxic Bt corn pollen to these rare butterflies. She crafted a petition to the EPA that would require farmers to plant a buffer zone of non-Bt corn around Bt cornfields as a way to reduce the amount of toxic pollen blowing into the habitats of butterflies. The petition recommended that these non-Bt corn zones be at least sixty to eighty feet wide, or about sixteen to thirty-two corn rows deep.

At the end of May 1999, farmers across the United States had just finished planting about twenty-seven million acres, accounting for 35 percent of the nation's corn crop, with biotech seed. Most of it was engineered to produce Bt toxins. At the St. Louis headquarters of the National Corn Growers Association, a group known for its vigorous support of genetic engineering, the phones started ringing. "We're hearing from farmers who are concerned about the perception of the corn pollen being toxic," Kevin Aandahl, who was then the group's spokesman, told me in an interview soon after the *Nature* article appeared. "We are concerned about the implications of the study, but we're urging caution of those who might leap to the conclusion that genetic crops are inherently bad for nature," he said. Aandahl was quick to add that John Losey himself had said that Bt corn does not harm humans or mammals.

Biotechnology industry officials downplayed the importance of the study, characterizing it as merely one laboratory experiment with little or no relevance to monarch butterflies in the wild. Val Giddings, vice president of BIO and a former USDA official, immediately turned out a press release stating that it was "highly likely that in the natural setting, outside the laboratory, most monarch larvae would never encounter any significant amounts of corn pollen."

"Timing is all important," Warren Stevens, senior curator of the Missouri Botanical Garden in St. Louis, was quoted as saying, in the same BIO press release. "For any harm to occur, the monarch larvae have to be emerging and feeding at the same time corn is pollinating, a narrow period of seven to ten days. Pollen has to be on a particular leaf that the caterpillar is feeding on. And the caterpillar has to consume the pollen before rain or dew washes it away."

Chris DiFonzo, a Michigan State University field crops entomologist, asserted that "Bt corn is a much safer method of pest management, and has less detrimental impact on all aspects of the environment—monarchs included—than the use of broad-spectrum insecticides."

Many other experts shared DiFonzo's belief that Bt crops would be safer for the environment, on balance, than conventional crops treated with chemical pesticides. In fact, the entire rationale for creating Bt crops rested on the premise that genetically engineered plants would significantly reduce the amount of chemicals applied to farmlands. When conventional pesticides are sprayed on farm fields, only a small fraction of the active ingredients is taken up by the weeds, insects, and crops. The bulk of the pesticides soak into the soil and leach into the groundwater, often polluting aquifers that flow deep beneath the Earth's surface. Pesticides trickle from farmlands into ditches and creeks and make their way to rivers and lakes, where they accumulate in fish. Contamination of drinking water wells by nitrates, insecticides, and weed-killers such as atrazine is a well-documented health threat across farm states. Nor is the pollution from chemical pesticides confined to farming regions. Persistent organochlorine pesticides applied to fields in one part of the world are carried on water vapor to distant locations, where they rain down on land and lakes.

Monsanto, Rhône-Poulenc, and DuPont, companies that made fortunes manufacturing pesticides, now promoted Bt crops as an environmentally friendly pest-control alternative to their own chemical products. The EPA embraced the development of Bt cotton, corn, and potatoes as a way to cut pesticide use. It encouraged Monsanto and other biotech companies to produce Bt plants by categorizing them as biopesticides, which are deemed an inherently safer class of pesticides than conventional chemicals.

Until John Losey's monarch study was published, most American environmental groups, too, had accepted the premise that genetically engineered crops would reduce the enormous load of toxic chemicals applied to the land. As long as the new crops caused no apparent harm—and quite possibly carried environmental benefits—the major conservation organizations took no position on their burgeoning use. The domestic critique of the biotech industry had been mounted primarily by two scientists at the Union of Concerned Scientists, a scientist at the Environmental Defense Fund, the organic farming community, and a handful of small grassroots consumer groups.

But the specter of seemingly benign corn pollen killing the nation's most cherished butterfly jolted major American environmental groups into action. Americans could no longer assume, as the government, university scientists, and biotech industry had repeatedly stated, that genetically engineered crops were better for the environment than other approaches to agriculture. The question of whether the widespread cultivation of Bt corn actually did decrease the amount of pesticides sprayed on farmlands took on great importance. But because neither the EPA nor the USDA required biotech companies to perform farm-scale studies investigating the ecological risks and benefits of gene-altered crops, there were few hard facts to debate.

More than a year after John Losey published his monarch butterfly study, an article on the risks and benefits of genetically engineered crops appeared in the December 15, 2000, issue of *Science,* concluding, "A review of existing scientific literature reveals that key experiments on both the environmental risks and benefits are lacking. The complexity of ecological systems," authors L. L. Wolfenbarger and P. R. Phifer stated, "presents considerable challenges for experiments to assess the risks and benefits and inevitable uncertainties of genetically engineered plants."

Before the advent of Bt corn in 1996, very few farmers bothered to apply pesticides to their corn crops to protect against European corn borers. The pests were known to take a big bite out of corn production—U.S. yield losses from the European corn borer averaged an estimated $1.2 billion a year. But scouting for corn borers is labor-intensive work, and historically most farmers chose simply to ignore the pests and accept the yield loss. Only about seven percent of American farmers sprayed their fields for European corn borers. Some studies showed a small drop in the amount of pesticides sprayed on the nation's corn crop in 1998 and 1999, but the dip in synthetic chemical use was accompanied by a sharp increase in plantings of hybrids engineered to make their own toxin.

BIO's Val Giddings also asserted that monarch butterflies faced a greater threat from people driving cars into them than from Bt corn pollen. "It's not an exaggeration to say more monarchs succumb to high-velocity collisions with car windshields than ever encounter corn pollen," he said.

During the fall monarch migration, it is not uncommon for motorists along the eastern shore and the midwestern flyways to encounter thousands of monarch butterflies moving south, some skimming just a foot or two above the surface of the road. I asked John Losey what he thought about Val Giddings's claim that monarchs faced more danger from automobiles than from Bt corn pollen.

"I don't think that's true, although I haven't seen any data—and I haven't seen anyone who quotes that [assertion] report any data. Even if it is true, it's important to think about the implications, because there were as many monarchs dying on windshields before the advent of Bt corn five years ago as there are now. The question is, you had that mortality before Bt corn and you still have it now, so that mortality has stayed steady, and now we're introducing this new effect," Losey responded.

"Say that windshields are taking out ninety percent of the monarch population," he continued. "Clearly, for the last thirty years, monarchs have survived that, but you could have ninety percent mortality from that, add three percent from the Bt corn, and maybe ninety-three percent might be enough—at some level they couldn't survive it. So that statement doesn't make any sense. If you think about it in a logical, quantitative, or scientific way, that statement has no merit."

While scientists debated the finer points of relative risks and benefits from gene-altered crops, most laypersons I talked to who had read about Losey's study experienced swift negative reactions to the high-tech corn. The image of thick clouds of lethal corn pollen blowing across the American heartland in the summertime reinvigorated the European epithet *Frankenfood*. If scientists at the EPA and the USDA had failed to predict the harmful effect on North America's favorite butterfly, what other unpleasant surprises lay down the road? Would honey become tainted? Would some insects become sick from eating the toxic corn pollen? Would the birds and animals that prey on insects that ingest Bt pollen themselves become weakened?

Throughout the summer of 1999, unanswered questions about Bt corn pollen fueled a growing public uneasiness about gene-altered crops. As the U.S. media played catch-up, each week brought more magazine features, newspaper editorials, and news stories about the environmental effects of Bt corn. Consumer activists and environmental groups besieged USDA, EPA, and FDA officials with questions about the health and safety of genetically engineered foods.

In late June a rumor began circulating in Washington that Secretary of Agriculture Dan Glickman was getting ready to ask the agribusiness and food industries to label genetically engineered corn and soybeans and separate them from conventional commodities. It wasn't clear, though, that Glickman had the authority to require such labeling. The FDA has jurisdiction over most food labels, with the exception of meats. With pressure mounting on the government to do something to shore up public confidence in the new crops, the USDA announced that Secretary Glickman would give a major public policy address on genetically engineered food at the National Press Club on July 13.

At the outset of his tenure as agriculture secretary, Dan Glickman had been an enthusiastic supporter of gene-altered crops. The former Kansas congressman had led the Clinton administration's fight to force Europeans to accept America's unlabeled gene-altered corn and soybeans. He sometimes began speeches by recounting an incident that befell him in Italy in 1996: as he was giving an address in Rome, naked protesters suddenly entered the hall and hurled genetically engineered soybeans at him. Activists notwithstanding, he would assure audiences that biotechnology would be an important tool in mankind's ongoing battle against hunger. In the spring of 1999, though, traces of uncertainty began to creep into his speeches. He started hinting *sotto voce* that the Clinton administration might not be able to force biotech crops down the world's gullet.

In a speech delivered in Washington, D.C., on March 19, 1999, as part of the USDA's Millennium Speaker Series, Glickman reaffirmed the administration's intention to use the World Trade Organization as a vehicle to open export markets for American biotech crops. "We must continue to argue in multilateral forums like the WTO that our biotech products have withstood the strictest scientific scrutiny," he said in the address. However, he warned, "market access isn't enough if, when it comes right down to it, many European consumers fundamentally don't trust and won't buy the products."

Glickman also argued that the USDA has an obligation to protect the rights of organic producers and consumers as well as conventional growers. Organic farming represents only a tiny percentage of U.S. food production, but it is one of the fastest-growing segments of the food industry. In 1997 organic food sales in the United States were about $4 billion, according to the Organic Trade Association. By 1999 sales of organic foods were estimated at $6 billion. Typically, organic foods command a premium price on the market over conventionally grown foods, often by as much as 20 percent.

During the 1980s, when organic farmers were developing systems for raising pesticide-free foods, several certifying organizations had formed to ensure consumers that foods labeled "organic" had met specific production standards. In 1998 the USDA proposed organic farming rules that would provide the industry with a uniform set of national standards, instead of the hodgepodge of regulations that existed from state to state. The USDA's draft proposal contained a provision allowing both genetically engineered crops and irradiated foods to be labeled "organic." Response to the proposal from consumers across the nation was swift and overwhelmingly negative. The USDA received hundreds of thousands of comments from citizens opposed to genetically modified foods.

"Let's remember," Glickman said in his millennium address, "that there are people who have other food preferences, and USDA must be responsive to them. Last year we heard from 280,000 organic consumers who do not want any genetically modified organisms in their food," he reminded his audience. "Are we at USDA doing enough to serve that market? As we discuss biotechnology issues, are we giving adequate consideration to biodiversity as well?"

During a speech at Purdue University in West Lafayette, Indiana, the following month, Glickman stunned American agribusiness and research institutions by openly stating that the government could not compel citizens in other countries to import biotech foods. The United States "can't force these new genetically engineered food products down consumers' throats," the secretary stated on April 29, 1999. He warned that "dismissing the skepticism that's out there is not only arrogant, it's also a bad business strategy."

This commonsense analysis angered many in the agribusiness community. American farmers had adopted Monsanto's new biotech seeds with the blessing, and often at the urging, of the USDA. Through its pervasive network of experimental field stations, land grant university research departments, and extension programs, the department exerts tremendous influence over the planting choices of conventional farmers. The USDA had assured growers the new biotech crops were safe and beneficial for the environment. During the summer of 1999, conventional farmers in the corn belt seemed largely dismissive of the doubts that European consumers were raising about the safety of biotech foods. When asked if he was at all worried about possible health or environmental problems from gene-altered crops, one midsize corn and soybean grower in Garden City, Iowa, responded in words that reflected a viewpoint widely shared in farm country: "If USDA says they're safe, they're safe."

In his April 29 speech, Secretary Glickman added, "My confidence in biotech—or the industry's confidence in biotech—is ultimately irrelevant. Only when consumers have confidence, and when they express that confidence at the grocery-store checkout line, will we be able to see the return on the enormous public and private investments we've made in biotechnology."

Glickman concluded on a warning note: "We have a way in this country of latching on to solutions, pursuing them to the exclusion of others, and then watching them sometimes backfire. So yes, let's be enthusiastic about these technologies and pursue them. But let's not put all of our eggs in the biotech basket."

The secretary's speech came one day after the British subsidiary of Nestlé announced on April 28, 1999, that it would attempt to eliminate genetically modified ingredients from all of its food products. A day earlier, on April 27, two British units of Unilever had declared their intent to stop using those ingredients as well.

The U.S. food industry was starting to grow nervous that American consumers would follow the lead of Europeans and reject genetically engineered foods. The Grocery Manufacturers of America quickly put together a public relations effort aimed at heading off a consumer backlash. In June 1999 the trade group made a special $1 million assessment of its members to create the Alliance for Better Foods. The alliance, which one GMA official called a "virtual organization," exists mostly on the Web. "In today's world, a million dollars or so is a very modest amount, so we didn't have enough money to do much more than inform policymakers such as members of Congress and regulators, and opinion leaders, folks who might be thought of as leaders in their community, and folks in the media—just inform them about the benefits of biotech," GMA vice president Gene Grabowski explained. The alliance focused on putting together media kits for reporters and paying special visits to members of Congress.

The food industry viewed the alliance as a stopgap measure, while the biotechnology industry developed a full-blown public relations program. That effort, run by a group called the Council for Biotechnology Information, was launched a few months after the alliance with a budget of $50 million. One of its products was a widely disseminated television ad featuring the slogan: "Biotechnology: a big word that means hope."

The Alliance for Better Foods represents thirty-five of the nation's top food and agriculture trade groups, including the American Bakers Association, the American Meat Institute, the National Soft Drink Association, and the National Restaurant Association. The alliance credo states: "With biotechnology producers can provide a more abundant, better-quality, and more nutritious food supply to consumers."

Even as American food companies were touting the benefits of genetically modified foods to domestic consumers, however, they were producing GMO-free foods for their European customers. When mandatory labeling laws took effect in Europe in 1998, many U.S. food manufacturers quietly changed the formulation of foods destined for EU markets. Rather than risk losing sales to competitors by listing unpopular GM ingredients on a label, most American companies, like their UK counterparts, were attempting to obtain nonengineered corn and soybeans for foods sold abroad. By the spring of 1999, few foods were to be found on British grocery store shelves bearing a label announcing genetically modified ingredients.

The GMA's director of international and environmental affairs acknowledged, at the annual meeting of the American Agri-Women in Washington in June 1999, that American companies were complying with European food laws that required labels on genetically modified foods. But the official insisted that identifying those foods on a label would not convey any "meaningful information" to American customers.

GMA officials have devoted significant resources to fighting state laws that would require labeling of genetically engineered foods. Matt Stanard, manager of state affairs for the GMA, delivered testimony in

Maine on April 12, 1999, against a proposed bill mandating such labels, arguing that it was misguided and unnecessary. "Following years of testing, the FDA has determined that biotechnology-enhanced foods are equivalent to foods developed through crossbreeding and traditional methods," Stanard said. "Thus, compulsory labeling provides no significant or useful information to consumers."

Monsanto too had abandoned its initial position that segregation of genetically modified commodities from conventional commodities was impossible. In April 1998 it formally reversed its policy against labeling gene-altered products bound for European markets. Tom McDermott, Monsanto's spokesman in Brussels, said the company had come to realize that in Europe a label was required, to make consumers comfortable with biotech foods.

McDermott told me that Monsanto was still philosophically opposed to labeling. The logic was seriously flawed, he argued, because once a society agrees that "process" has validity, there will be no end to the kinds of information that people will want to see on food labels. He characterized Monsanto's new EU policy as a pragmatic approach to the reality that European consumers want to know when they are purchasing genetically modified foods. But Monsanto did not extend its labeling policy to domestic consumers.

By the spring of 1999, labeling had become a flashpoint for the American consumer resistance to genetically modified foods. In late March 1999 Craig Winters of Seattle founded the Campaign to Label Genetically Engineered Foods. Winters, who is an Internet-based activist, named Alexander Schauss as president of the campaign's board of directors and Marlene Beadle as treasurer. Schauss, Beadle, and Winters had formerly worked together in 1992, when they created an activist group called Citizens for Health that helped lead the successful legislative effort to pass the Dietary Supplement Health and Education Act of 1994. The intent of that activist campaign was to give consumers freer access to vitamins and herbal products from traditional medicine. The 1994 law decreased the FDA's authority to restrict sales of dietary supplements. Winters set as his goal getting Congress to pass legislation requiring labels on all genetically engineered foods sold in the United States.

Winters picked up the labeling effort where another small grassroots group had left off. In 1997 Mothers for Natural Law had started an education campaign to raise public awareness about the genetic alteration of America's food. Susan Marcus, an Iowa high school teacher and mother of two girls, started volunteering for Mothers for Natural Law shortly after she learned about the controversy in Europe and Japan. She had called the FDA to find out what kinds of safety testing the U.S. government was doing on genetically engineered foods. "A spokesman there told me that the FDA was doing very little testing of its own," Marcus said, "because their research methods wouldn't work for genetically engineered food. He told me that they were leaving it primarily to the biotech companies to decide for themselves when their products were safe for the market. I felt that it wasn't right that I couldn't choose whether to feed these foods to my family or not. They're on the market. They're not labeled."

On June 4, 1999, the Campaign to Label Genetically Engineered Foods wrote to EPA administrator Carol Browner seeking an immediate moratorium on the planting of genetically engineered crops until sufficient studies to determine their safety had been completed. Two weeks later Mothers for Natural Law presented members of Congress with nearly half a million signatures from Americans demanding labeling of genetically engineered foods.

The major biotech policy speech that Secretary Glickman delivered on July 13, 1999, left environmental and consumer groups bitterly disappointed. Glickman ordered an independent review of the USDA's system for approving new genetically engineered crops, a move welcomed by the environmental community, but he failed to call for mandatory labels on the foods.

The secretary acknowledged that genetically engineered food technologies are "in their relative infancy," and warned that society "cannot blindly embrace their benefits." He asked developers of bioengineered products to report "any unexpected or potentially adverse effects" to the USDA immediately upon discovery.

Glickman also vowed that government regulators would "stay at arm's length" from the companies whose genetically engineered products they had approved. This pledge was intended to blunt domestic and European critics who had long assailed the cozy relationship between the USDA and the biotechnology companies. Through campaign contributions and masterful manipulation of the legal revolving door in Washington, Monsanto had succeeded for years in influencing, and often shaping, the government's biotech policy.

Perhaps the most egregious and widely reported instance of the cozy relationship between Monsanto and its government regulators is exemplified in the work history of Michael R. Taylor. Taylor joined the FDA in 1980, serving for four years as executive assistant to the commissioner. As Peter Montague exposed in a March 1994 article in the weekly environmental newsletter *Rachel's Hazardous Waste News,* Taylor subsequently joined the Washington, D.C., law firm of King & Spaulding, where he worked from 1984 to 1991. During that time, Monsanto sought FDA approval for its genetically engineered bovine growth hormone. In 1991 FDA commissioner David Kessler brought Taylor back into the agency to serve as deputy commissioner for policy. In that capacity, Taylor was involved in writing the FDA's rBGH labeling guidelines, which warned retailers not to label milk as "BGH-free." From 1994 to 1996 Taylor served as administrator of the USDA's Food Safety and Inspection Service. In September 1998 Monsanto announced that Taylor would become vice president for public policy and head of the company's Public Policy Institute.

Other prominent people who passed through the revolving door include Michael "Mickey" Kantor, a close personal friend of Bill Clinton and the president's first U.S. trade representative, who later became a member of Monsanto's board of directors, and Marcia Hale, who became assistant to President Clinton for intergovernmental affairs in 1993 and went on, after leaving the administration, to become Monsanto's director of international government affairs. In May 1995 President Clinton named Monsanto chairman and CEO Robert Shapiro to the nation's Advisory Committee for Trade and Policy Negotiations.

Monsanto culled dozens of lobbyists from the ranks of former White House and congressional staffs. In 1996 and 1997 more than twenty-five lobbyists for Monsanto who filed disclosure forms with the Senate and House of Representatives had worked previously as

congressional or White House staffers. Monsanto also contributed
generously to both Democratic and Republican lawmakers. Between
1987 and 1996 the company contributed $830,407 to congressional
campaigns, according to the Center for Public Integrity in Washington, D.C. During the same period, the American Crop Protection
Association, of which Monsanto is a member, contributed another
$137,800 to congressional races.

In response to the agriculture secretary's speech, the *Minneapolis
Star Tribune* ran an editorial on July 23, 1999, entitled "Franken-
foods: Doubts over Biotech Deserve Respect." The newspaper pointed
out that the USDA has "the conflicting tasks of promoting new biotech products while ensuring that they pose no harm to other plants or
animals." The editors voiced disappointment over what they called
Glickman's "wishy-washy position on the question of labeling genetically modified foods as such—a step demanded by some consumer
groups and emphatically resisted by industry." It would take years for
science to answer all the doubts and fears about genetically engineered
foods, the editors predicted, concluding that "in the meantime, consumers ought to be able to choose whether to eat the stuff."

Throughout the summer of 1999, opposition to genetically engineered
crops gathered strength among major national consumer and environmental groups. On August 11, 1999, the National Wildlife Federation, the Natural Resources Defense Council, the Arizona-Sonora
Desert Museum, the Union of Concerned Scientists, and the Defend-
ers of Wildlife petitioned EPA administrator Carol Browner to restrict
future plantings of Bt corn.

The following week the Sierra Club wrote a letter to President Clinton demanding more research into the risks that genetically engineered organisms pose to both humans and wildlife. With more than
seven hundred thousand members, the Sierra Club, founded in 1892,
is the largest conservation group in the United States. Carl Pope, the
Sierra Club's executive director, warned Clinton that the rate of application of the technology "far exceeds our ability to understand the
environmental and public health risks to avoid potentially serious
harm."

Members were concerned, Pope wrote, that pollinator species such as bees might be harmed by Bt crops, "with disastrous consequences to the food supply." The discovery of the potential deaths of monarch caterpillars from Bt toxin only after millions of acres of Bt corn were already in the ground was "a dramatic example of adverse secondary effects from a technology that is inadequately understood."

The Sierra Club called on the administration to "end the concept of 'substantial equivalence' by our regulatory agencies as a ploy to side-step safety studies and oversight responsibilities. For example, toxins meant to kill insects are being genetically engineered into plants, yet the consequences of these toxins in the diets of humans, livestock, beneficial insects, and wildlife are unknown."

Some U.S. activists took their concerns about genetically engineered foods straight to the food manufacturers. Shortly after John Losey's study was published, Charles Margulis of Greenpeace faxed a letter to the CEO of the Gerber baby food company in Fremont, Michigan, asking whether the company used genetically engineered ingredients in its products. *The Wall Street Journal* reported in a front-page story on July 30, 1999, that the letter got the immediate attention of the company's parent company, Novartis, which is headquartered in Switzerland. Gerber officials in Michigan apparently were not worried about American consumers learning that the company used gene-altered ingredients in baby food, but officials at the parent company's Swiss headquarters were very worried, the newspaper reported. The CEO of Novartis pressed for a speedy resolution of the problem. Gerber decided in July to remove all genetically engineered corn from its dry cereals and other baby foods. In addition to seeking organic corn suppliers, the company also said it would search for organic soy growers to provide ingredients for all of its baby food products.

Gerber's move stunned many in the U.S. food industry. The united show of support for genetically modified foods once displayed by GMA and the National Food Processors Association had begun to crack. Within days two other makers of baby food, H.J. Heinz Company and Healthy Time Natural Foods, followed in Gerber's footsteps, stating their intent to eliminate genetically modified ingredients from their products. A Heinz official told the Associated Press on August 1, 1999, that with the exception of foods sold in the United

States, most of its worldwide products were already "virtually free" of gene-altered ingredients.

On August 24, 1999, the Consumers' Union, the organization that publishes *Consumer Reports,* released a list of foods the group had bought in American grocery stores that tested positive for genetically modified ingredients. Three infant formulas that Consumers' Union submitted to an independent lab for testing contained genetically engineered soybeans: Enfamil ProSobee Soy Formula, Similac Isomil Soy Formula, and Nestlé Carnation Alsoy. Several soy burgers tested positive, including Boca Burger Chef, Max's Favorite, and Morningstar Farms Better 'n Burgers. Other popular processed foods that tested positive were Ovaltine Malt powdered beverage mix, Bacos bacon flavor bits, Bravos Tortilla Chips Nacho Nacho!, Old El Paso 12 Taco Shells, and Jiffy Corn Muffin mix.

None of these foods were labeled as containing genetically modified corn or soybeans, prompting the Consumers' Union to advocate mandatory labeling. Two physicians endorsed the recommendation.

"No long-term studies of the impact of genetically modified foods on health, particularly the health and development of babies, the sick or elderly, have been done," said Martha Herbert, a pediatric neurologist at Massachusetts General Hospital. "The promoters of this uncontrolled and profit-motivated experiment on people and nature have no scientific grounds for claiming these substances are safe."

Paul Billings, a physician affiliated with the Council for Responsible Genetics in Cambridge, Massachusetts, decried the lack of safety testing for gene-altered foods. "The U.S. government has extensive tests for new drugs developed through biotechnology," Dr. Billings said. "Biotech foods have the same properties, the same potential for adverse effects on health or the environment, but we don't test them. It's not responsible."

The September 1999 issue of *Consumer Reports* was careful to emphasize that no one had found any health problems linked to eating genetically engineered corn or soybeans: "There is no evidence that genetically engineered foods on the market are not safe to eat." But as Mark Silbergeld of the Consumers' Union told the Senate Agriculture

Committee the following month, "At the same time, neither can it be said that they have been proven 'safe.' "

Consumers' Union scientists were troubled by the lack of independent safety tests and long-term studies demonstrating the new foods' safety. Between safe and unsafe there is a "no man's land" where we don't know the long-term effects, Silbergeld argued. To say we know is "an expression of faith, not knowledge"—especially if what's being introduced is not from a food product.

Three organizations released a joint statement supporting *Consumer Reports'* assertion that there is no evidence showing that the foods are unsafe. "Crops enhanced through biotechnology have become commonplace in the U.S. years after health professional organizations such as the American Dietetic Association and the American Medical Association endorsed their safety and promise of future benefits," wrote the Georgetown University Center for Food and Nutrition Policy, the International Food Information Council, and the Council for Agricultural Science and Technology.

"We can be sure that biotech foods are safe, and I think that's very important," said Edith Hogan, a spokesperson for the American Dietetic Association. "It's important to me—as a mother and as a grandmother—as well as a dietitian, to feel secure about our food supply."

I called the American Medical Association (AMA) at the end of August 1999 to learn more about the basis for the prestigious group's endorsement of genetically engineered foods. A senior scientist in the AMA's unit on science, research, and technology told me that the AMA planned to review its policy because it was formulated back in 1990. "We recognize the policy is stale and needs to be revisited," the AMA official said. "It's the only policy we have, and I think it reflects the way things were ten years ago, but a lot of things have happened since then, with the British Medical Association coming out and saying they were revisiting the issue."

Some consumers found an outlet for their frustration over the government's refusal to label genetically engineered foods in a little-known forum: the World Trade Organization. Throughout the summer the

USDA held a series of "listening sessions" in rural communities across the nation. The purpose of the public meetings, the USDA said, was to help the government prepare for the upcoming WTO meeting that the Clinton administration was to host in Seattle at the end of November. At each listening session, USDA officials gave members of the public a chance to offer their opinions about the kind of agriculture and trade policies they believed would most benefit America. Many of the consumers who packed the public meetings voiced intense anger over the stealth appearance of genetically engineered ingredients in so much of the American food supply. Some of the growers who spoke also questioned the wisdom of planting biotech crops when the markets abroad were closing to the new foods.

USDA undersecretary Jim Schroeder held a public listening session at the Holiday Inn in St. Paul, Minnesota, on June 7, 1999. Many local growers told Schroeder they feared that markets for U.S. commodities were drying up overseas because of the overwhelming rejection of genetically modified crops in Europe and Japan.

Howard Fleager, a Minnesota grower representing a farmer-owned company called American Growers Food, described conversations his company had held with grain importers during recent trips to Japan. He learned that Japanese importers had no desire to purchase genetically engineered crops. "The Japanese people are scared to death of genetically modified organisms, or GMO crops. The most consistent comment that they made was, 'We will watch the children in the United States for the next ten years,'" Fleager told USDA officials.

"Let's make the GMO issue simple. Let's label all of it and let the consumer decide—whether it be here in America or abroad," Fleager added.

Steve Sprinkle, an organic farmer since 1976, said organic growers had become "accidental beneficiaries of all the maladies that are now befalling conventional farming." Many European consumers, particularly in Great Britain, he said, had turned to organic foods as a way to be certain they were not ingesting genetically modified ingredients. Major food companies such as Unilever and Nestlé had elected to keep gene-altered soy lecithin out of their candy bars and other processed foods.

Other growers, including Don Louwagie of Marshall, Minnesota, urged the government to use the WTO talks to open up markets for American farmers. They hoped the United States could use its diplomatic muscle in the upcoming negotiations to force Europe to accept genetically modified crops. "Gaining improved access to foreign markets is of critical importance to soybean farmers," Louwagie said. "Every other row of soybeans produced by growers is exported overseas in the form of soybeans, soybean meal, or soybean oil. Soybeans and soy products are our nation's largest agriculture export commodity. Rules governing biotech trade must be included in the next WTO round in order to ensure science-based regulatory reviews and trade rules. These WTO rules must supersede the rules of any other international treaty or agreement."

On July 19, 1999, during a WTO listening session in Burlington, Vermont, the USDA heard still more objections. Detlev Koepke, a history professor at Bridgewater State College in Massachusetts and the founder of a cooperative market, asked USDA officials for a "moratorium on U.S. support of genetically engineered foods until mandatory testing is done to assure the long-term safety of these foods, which has not been done, so that consumers can make informed decisions about whether to buy them or not." He warned that genetically engineered foods could turn out to be the thalidomide of the next decade.

"As the percentage of unlabeled GE foods spreads in American products, other countries will increase their bans," he predicted. "Even if WTO succeeds in forcing Europeans to accept American products, European and other consumers will refuse to buy them. The same reaction is beginning to happen in America as well, as consumers become more educated."

At the same listening session, a member of the public, Betsy Gentile, said that some very rich corporations in the United States were "trying to force bioengineered products on countries whose scientists warn them of some frightening possible health risks associated" with the foods. "Sometimes corporations do lose their way. Wealth doesn't guarantee wisdom, or even common sense. And now here we are again together with the chance to say no to those who want to test the products on all of us and our children," Gentile continued. "Food is not just a commodity. Food is holy, and I believe the small farmers of Vermont and the small farmers all over the world can benefit only from providing foods people know is healthy. Bioengineered foods have not been proven healthy. New studies every day raise more concerns."

Concluding her emotional statement, which drew applause from the Burlington audience, Gentile stated, "People should be able to refuse food that their best information leads them to believe might be poison. And I believe that if Dante was still writing, he would create a special circle of hell for those who ask us to experiment on our own children."

NINE

DYING ON THE VINE

I n the autumn of 1999, after a summer of setbacks, the biotechnology industry found itself reeling from the ferocity of the emerging backlash against genetically engineered food. Threats to the monarch butterfly had galvanized American environmental groups, organic farmers, and consumer organizations in a fierce struggle to stop the spread of the crops and to force safety testing of gene-altered foods already on the market. While activists were fired up by the monarch butterfly issue, however, industry polls revealed that the overwhelming majority of Americans still knew surprisingly little about genetically modified foods. That fall the food and biotechnology industries regrouped for a public relations offensive to win over the hearts and minds of American consumers.

Considering the situation in Europe, biotech food advocates feared they had little time.

In mid-September Joyce Nettleton, director of science communications for the Institute of Food Technologists (IFT), opened a meeting on international trade and biotech food with a foreboding question: Are GMO foods dead? She warned the food scientists, regulators, and academics gathered at the German Embassy in Washington for the IFT-sponsored meeting that unless science societies and the food industry managed to change the agenda quickly, the enormous potential of biotechnology would be placed in the deep-freeze. She urged those assembled not to concede the agenda to activists.

As the biotech industry meeting tried to understand what had gone wrong, recriminations flew between Americans and Europeans, industry officials, diplomats, and bureaucrats. Some company officials had canceled summer vacations to deal with the burgeoning crisis. An official with the American Crop Protection Association told me about a curious conversation she had had with a male colleague at the National Cotton Council of America. A woman who had telephoned was worried about possible health problems from tampons made with genetically engineered cotton. Neither of them had had the faintest idea how to respond to the query. Some European women had already questioned the safety of using gene-altered cotton for bandages, nappies, and feminine hygiene pads.

The German counselor for agriculture blamed the life sciences companies for pushing their products too hard and for disregarding the final marketplace: consumers. He felt that the regulatory system had relied too much on expertise provided by the biotechnology industry itself. To its own detriment, he argued, the industry was ignoring the need for long-term studies on human health effects and ecological safety.

Tassos Haniotis, a representative of the European Commission agriculture sector and a frequent speaker on the subject of genetically modified food and trade, told the audience that throughout Europe every nation was completely rethinking food policy. Europeans wanted government to cut agricultural output and to place more emphasis on the quality rather than the quantity of food produced. Europeans were willing to take some risks in life, he noted—they accepted the health risks that come with smoking and driving cars—but they were not willing to tolerate any risk from companies' manipulating the genetic makeup of food.

An official with the office of the U.S. Trade Representative reminded the international audience that at the turn of the twentieth century, the public had rejected pasteurized milk, but gradually pasteurization became universally accepted. She predicted that over time biotech food products could entirely replace conventional ones.

On Capitol Hill, Representative Nick Smith, chairman of the House Science Committee's subcommittee on basic research, tried to marshal support for genetic engineering. When he convened his hearing on October 5, 1999, his remarks reflected the glum state of the biotech food industry. Agricultural biotechnology, "which holds so much promise for so many, is in danger of dying on the vine before its full potential is realized," he began. "Campaigns of fear, sometimes intimidation, by activist groups, would deny the benefits of these powerful new tools."

Anthony Shelton, a professor in Cornell University's entomology department, was his star witness. Shelton's testimony was aimed at discrediting John Losey's monarch butterfly study. The junior faculty member's correspondence in *Nature,* Shelton began, "has attracted considerable attention in the press and led to dramatic shifts in policies, including possible trade restrictions by Japan, freezes on the approval process of Bt-transgenic corn by the European Commission, and calls for a moratorium on further planting of Bt corn in the United States. Was this reaction justified based on what can only be considered a preliminary laboratory study? Absolutely not. Serious questions have been raised about the science of this study, but once it was published it became a rallying cry against Bt plants."

There was actually little dispute in the scientific community about the validity of Losey's finding that Bt corn pollen kills monarch caterpillars. In fact, several scientists had begun additional studies to expand upon Losey's finding. Some researchers had conducted tests that showed that Event 176, a variety of Bt corn made by Novartis, produced much higher—and more lethal—concentrations of toxin than other Bt corn hybrids. Other researchers had tallied the number of Bt pollen grains that a monarch caterpillar needs to eat to get a lethal dose. In the coming years, dozens more studies would nail down the particulars of Losey's initial finding as well, but in the fall of 1999, and later in 2001, no one quarreled with the fact that Bt corn pollen is deadly to monarch caterpillars that ingest enough of it.

One key outstanding question in need of an answer was, how much Bt pollen would monarch caterpillars encounter in natural settings? Laura Hansen and John Obrycki, entomologists at Iowa State University in Ames, Iowa, had spent three years studying the effect of Bt toxins on monarch caterpillars. They confirmed that Bt corn pollen spreads to the milkweed plants near cornfields, and found that within two days of ingesting the toxic pollen, 20 percent of monarch cater-pillars were dead.

One of the most vocal academic scientists to question the validity of Losey's study was Shelton himself. Representative Smith asked him whether Losey had jumped the gun, somehow publishing the results of his study too soon. "Do we have a system that is maybe overanx-ious on publishing something? Why are we doing the kind of studies that would be so farfetched from nature in terms of the lab study, for example, on the Bt and the monarch?" Smith asked. "Are we too anx-ious to study, or to publish, prematurely?"

Shelton, a senior member of Losey's department, responded, "I don't think it was a realistic study."

Smith seemed perturbed by the notion that a federally funded scien-tist would be allowed to undertake a study that might turn up possible problems with gene-altered corn. "Help me understand. Was this a taxpayer-funded study?" the Michigan Republican pressed.

"It was actually not supported on any federal funded grant that I know of," Shelton rejoined. "And in fact, I reviewed the paper before it was submitted. It was submitted to *Science,* and I said, 'I don't think you have a story here.' It was actually rejected by *Nature,* but *Nature* decided that they would take it as a news brief. And I would also fault the sensationalism of some of the scientific publishers for contributing to this as well."

Shelton insinuated that it was irresponsible of Losey to publish his study results when he did. "I think also, as scientists, we have an in-credible responsibility to act in a reasonable fashion, and there is a lot of pressure on scientists to sometimes get things out before they really should be," he told the committee. "And I think everyone in a univer-sity has to make sure that their colleagues act responsibly, and also, that if we can, realize the repercussions if we don't."

Shelton's veiled suggestion that university scientists ought to look over their colleagues' shoulders seemed to imperil the old-fashioned image of academic researchers as unfettered, independent thinkers. For his part, John Losey fervently defends his decision to publish his findings when he did. "We found this data that there could be an ef-fect. In retrospect, about twelve labs across the country have re-searched it, and we still don't have a full answer. So I think to have sat on the data ourselves and not publish it when we did would have been irresponsible and unethical," Losey argued when I interviewed him in 2000. "When you have something like that, I don't think it's right to just keep it in your own lab and not publish it. You've got to put it out there, so your scientific peers can all bring their expertise to bear on it, and so the public can have their discussion and debate on it. That's why I bristle a little bit when people say, 'You should have waited till there was more study done before you published it.' I really strongly disagree with that."

On October 6, 1999, Senator Richard Lugar, chairman of the Agriculture, Nutrition, and Forestry Committee, called into session a two-day hearing on genetically modified foods. Like his colleague in the House of Representatives, Lugar was distraught by the erosion of support; the federal government, corporations, public and private universities, and states had sunk billions of dollars into research and development of those foods.

"The biotech ball game may be over very rapidly," he warned. "As you take a look at the rest of the world, it's not a pretty picture."

The Indiana Republican pressed Clinton administration officials to be more outspoken in vouching for the safety of genetically modified foods. He told top officials from the USDA, the EPA, and the FDA that they had an obligation to rebut false statements made about those foods in the press. Industry wants a stronger seal of approval, the senator told the agency officials who testified on October 7.

To explain the dramatic nutritional changes genetic engineers might soon bring about in food crops, Lugar called Dean DellaPenna, a biochemistry professor at the University of Nevada in Reno, as a witness. DellaPenna is a pioneering researcher in a new field called "nutritional genomics," which, as he describes it, is an approach for discovering the genes that control micronutrient pathways in plants. While macronutrients consist of the carbohydrates, fats, and proteins in foods, the essential micronutrients include seventeen minerals and thirteen vitamins. DellaPenna was exploring ways to increase the levels of micronutrients in food plants.

Over 30 percent of the world's population, he told the committee, suffer from one or more severe deficiencies in micronutrients, with women, children, and the elderly at special risk. "Even in the industrialized nations, micronutrient deficiencies are surprisingly common, due to poor eating habits," he remarked. In the United States 80 percent of teenage women take in less than the recommended daily allowance for iron. Half the population takes in too little calcium, and 30 percent get too little vitamin E.

DellaPenna said studies have shown that vitamin E supplements can reduce the risk of coronary heart disease by 40 to 50 percent, and the risk of certain cancers by 70 percent. "One way to ensure an adequate dietary intake of essential micronutrients would be to manipulate their levels in plant foods," he told the Senate committee. His laboratory had isolated a gene from a simple nonfood plant that could boost the production of the potent form of vitamin E dramatically. He hoped to "move the technology" into agricultural crops such as soybeans, corn, and canola.

DellaPenna had acknowledged elsewhere that this approach to human nutrition was not without risks. Researchers would have to be careful lest gene-altered crops deliver more than the maximum safe dietary level of certain nutrients. Too much iron or vitamin A, for instance, can cause serious adverse health effects in people. If a woman takes excess vitamin A during pregnancy, the chances are increased that her baby will have birth defects.

Gary Kushner, counsel for the Grocery Manufacturers of America, told the committee that Clinton administration officials needed to speak out more forcefully in "support of the science-based regulatory systems in place in the United States that assure consumers approved biotech commodities are safe and wholesome." Echoing Senator Lugar's sentiments, he exhorted food companies, lawmakers, scientists, and regulators to make sure that "activists with a political agenda do not kill the promise of biotech foods."

Despite the growing public clamor for labels on genetically modified foods, the GMA's Kushner argued strenuously against mandatory labeling. "If the U.S. government were, in any way, to change its policy and require special labeling for biotech foods, such labeling could have the effect of misleading consumers into believing that biotech foods are either different from conventional foods or present a risk or a potential risk—even though FDA has determined that the biotech food is safe," he warned the committee. He added that "special labeling of biotech foods could lead to the very kind of consumer confusion that labels are designed to prevent."

As House and Senate leaders were trumpeting the future benefits of genetically engineered crops, the nation's biotech bellwether was pulling in its clapper. Buckling to pressure from environmentalists and shareholders alike, Monsanto chairman and CEO Robert Shapiro held a conciliatory meeting with Greenpeace on October 6, 1999. In a high-tech encounter with his detractors, Shapiro's image was beamed by satellite to London, where Greenpeace International was holding its fourth annual conference for business leaders. A haunted-looking Shapiro appeared on a video screen, offering a mea culpa to the environmental group.

"I think we've tended to see our task to convince people that this is a good, useful technology; to convince people, in short, that we are right and that by extension people who have different points of view are wrong—or at best misguided," Shapiro told Lord Peter Melchett, an avid organic farmer and leader of Greenpeace's antibiotechnology campaign. "The unintended result of that has been that we have probably irritated and antagonized more people than we have persuaded. Because we thought it was our job to persuade, too often we forgot to listen."

In truth, Monsanto had little choice but to listen. The financial worth of the once-mighty multinational corporation was plummeting. Shapiro's aggressive quest for dominance in the plant biotechnology business had been a costly strategy. As opposition to Monsanto's gene-altered crops spread from Europe to the developing world and, after the publication of John Losey's butterfly study, to the United States, Monsanto's stock had suffered on Wall Street. The stock price dropped from a high of more than $62 a share in August 1998 to less than $42 a share in August 1999, where it hovered throughout the fall.

Shapiro offered to engage in a dialogue with consumers and environmentalists. He acknowledged that genetically engineered crops pose potential concerns as well as potential benefits. He also offered to reconsider the moral, religious, and ethical ramifications of the use of biotechnology in agriculture.

Shapiro's widely reported teleconference with Greenpeace followed by just two days his dramatic declaration that Monsanto would not commercialize the Terminator technology.

His decision to reject use of the seed sterilization technology, Shapiro said, was prompted by a request from Gordon Conway, president of the Rockefeller Foundation in New York. Conway, an ecologist by training, has long viewed agricultural biotechnology as a potent tool to improve the livelihood of farmers in the developing world. Conway believes that food plants engineered to resist drought, grow in poor soils, and contain more nutrients could foster a "doubly green revolution." In the decade of the 1990s, the Rockefeller Foundation spent $500 million on agriculture research for the poor, with about $100 million going to training scientists in biotechnology. Scientists from Korea, China, India, the Philippines, Thailand, and Vietnam traveled to the United States under the foundation's aegis to study genetic engineering and then bring the techniques back to their own countries.

Conway felt that Monsanto's aggressive marketing methods, coupled with the company's intended acquisition of Terminator technology, were generating bad press and ill will toward all applications of agricultural biotechnology. Poor countries, which might stand to benefit most from the kinds of crops Conway envisioned, were growing skittish about genetically modified food. In June 1999 Conway addressed the Monsanto board of directors meeting at a hotel in Washington, D.C., to express his misgivings about the seed sterilization technology.

Shapiro wrote a letter to Conway on October 4, 1999, disavowing development of Terminator technology: "I am writing to let you know that we are making a public commitment not to commercialize sterile seed technologies, such as the one dubbed 'Terminator.' We are doing this based on input from you and a wide range of other experts and stakeholders, including our very important grower constituency."

A month later the biotech industry faced a new challenge from an unexpected quarter. On November 10, 1999, four congressmen called a press conference in the Cannon House Office Building to unveil a historic piece of legislation, the Genetically Engineered Food Right to

Know Act. The stately room was jammed with reporters and camera equipment and its polished antique mahogany tables piled high with colorful boxes of name-brand foods—Aunt Jemima pancake mix, Heinz mixed-grain baby cereal, Duncan Hines fudge marble cake mix, Ovaltine, and Quaker chewy chocolate chip granola bars.

Ohio representative Dennis Kucinich strode to a podium behind the boxes of food. "The rapid emergence of genetically engineered foods into our nation's food supply requires us to be confident and vigilant that every health, safety, and environmental concern has been fully examined," he told reporters. "I'm not confident that all the negative impacts from introducing these new foods into our food supply have been detected and avoided, so I'm here today with my colleagues to help ensure the safety of our food supply."

Bucking the food industry and agribusiness interests in his own midwestern state, the fiesty Democratic congressman promised that his bill would require clear labels on all foods containing genetically engineered ingredients. "I firmly believe that American citizens must have the right to choose what foods they and their family eat. In the absence of a genetically engineered label, they will not have this choice," he exhorted.

Following Kucinich at the podium, Representative Peter DeFazio, an Oregon Democrat, gestured to a familiar box of Aunt Jemima pancake mix. "This box really hasn't changed much since I was a kid," he said. "Guess what? There is a big change, just a landmark change in the contents of this box in my lifetime. Some of this comes from soy which has been genetically modified." DeFazio charged that the "FDA has turned a blind eye to the genetic alteration of our food supply and the potential health effects."

DeFazio accused agricultural biotechnology companies of putting profits ahead of food safety and the environment. "They've pioneered this, they've invested tens of billions of dollars in this research, they're about to reap these huge, patented lucrative rewards, and oh my God, now the consumers want to know, and they might not buy the product!" he charged. "That's what this is all about. It's about profits."

Representative Jack Metcalf, a Republican from Washington, was among the bill's nineteen cosponsors. "I believe that bioengineered foods hold great potential for benefit to the consumer. However, I also believe that the American citizens should have the right to know what they are eating," Metcalf said, noting that the bill simply allows for "some commonsense labeling requirements."

Representative Bernie Sanders, an Independent from Vermont, warned that there were many "unanswered questions about these products that science is still in the process of answering. During the time it takes to learn more about these products, we need to at the very least label GMO products so that consumers know what is in the food they are eating." He added, "There is concern all over the world about this issue."

John Losey's finding that Bt corn pollen kills monarch butterfly caterpillars cast doubt on the validity of the EPA's claims that biotech crops are safe for wildlife and the ecosystem. EPA officials acknowledged that they did not require Monsanto and other biotech companies to perform the kinds of long-term ecological effects studies that some environmentalists had called for before allowing them to sell their patented new seeds.

Margaret Mellon of the Union of Concerned Scientists believed that the USDA and the EPA had approved industry applications for gene-altered crops without first examining the potential negative impacts to organisms on farmlands and the surrounding ecosystem. The American approach to the potential ecological hazards of agricultural biotechnology, she told me, could be summed up as "Don't look, don't find."

EPA scientists did ask Novartis, Monsanto, and other makers of Bt corn to perform studies to make sure that their new varieties would not harm several sentinel species, including green lacewings, honeybee larvae, honeybee adults, parasitic hymenoptera, and ladybird beetles. Perhaps not surprisingly, Monsanto, Novartis, and the others all submitted studies to the EPA that showed their products to be safe for bees, lacewings, and other wildlife species. EPA scientists who reviewed the industry test results backed their findings on the safety of the Bt corn toxin.

In 1999 a panel of scientists assembled by the National Academy of Sciences reviewed some of the industry studies. Their independent review turned up conspicuous problems with some of Monsanto's studies.

The academy found that many of the studies Monsanto performed by design would have been unlikely to expose possible harmful effects of Bt crops. In studies Monsanto submitted to the EPA, for example, thirty green lacewings were fed moth eggs that had been coated with Bt toxin. Monsanto concluded that there was "no difference in mortality or lower mortality associated with Bt treatment." But NAS reviewers noticed that the company had coated the outside shells of the eggs with the Bt toxin. Since lacewings typically feed on only the internal content of the eggs, the NAS concluded that the lacewings "may not have ingested much of the toxin."

The NAS review also uncovered faults with the methods Monsanto used to test the safety of the Bt toxin in genetically engineered potatoes. The academy was able to get more detailed information from Monsanto than is available to the public. When the independent reviewers looked at Monsanto's tests of Bt toxins on honeybee larvae, they said the study approach was so "inefficient" that there was no way to tell how much, if any, of the Bt toxin the larvae had actually ingested.

The reviewers also found shortcomings with Monsanto's ladybug studies. Ladybugs—or ladybird beetles, as they are sometimes called—are welcomed by gardeners because they eat aphids and other pests that feed on the stems and foliage of many cultivated plants. In the tests Monsanto submitted to the EPA, adult ladybird beetles were given a honey solution containing the Bt toxin. But there was no way to tell how much exposure the ladybugs had to the toxin, the NAS reported, "because there was no control for evaporative loss" of the solution.

The NAS report confirmed the fears of many environmental activists: the ecological experiment under way across millions of acres of America's heartland was based on flimsy industry studies. These little-understood Bt crops, activists went on to charge, might threaten not just majestic butterflies but a multitude of other beneficial insects large and small as well.

Janet Andersen, director of the EPA Biopesticides Division, which gave Bt corn the go-ahead, and other EPA officials did not seem worried, despite Losey's study, that genetically engineered corn would harm monarch butterflies in the wild. But EPA was under terrific pressure from activist groups to revoke Bt corn licenses or severely restrict acreage to be planted in the future. To justify continued use of Bt corn, the Biopesticides Division needed hard scientific data proving that it was safe. If the biotech companies wanted to protect their enormous investment, they had to quickly provide studies showing that Bt corn poses little danger to the nation's most popular butterfly.

Within weeks of the publication of Losey's study in May 1999, Monsanto, Pioneer, Novartis, Mycogen, and Aventis—the five companies that make Bt corn—scrambled to put together a crash research program. They needed to find out how much toxic Bt corn pollen monarch caterpillars are exposed to under natural conditions across the North American corn belt.

The companies set up a $100,000 kitty and contracted with Richard Hellmich, a researcher with the USDA's Agricultural Research Service at Iowa State University in Ames, to coordinate scientific studies during the growing season already well under way. There were a host of questions to be answered: How prevalent are milkweed plants around cornfields? How far does corn pollen migrate from the edge of a field? How many monarchs are produced in regions of the country where corn is grown? Do monarch caterpillars feed on milkweeds at the same time Bt corn pollen is being shed? What dose of Bt pollen is needed to kill a monarch caterpillar? What happens to monarch caterpillars that eat a sublethal dose of Bt pollen? Would their fitness for survival be reduced?

Scientific studies generally require careful advance planning, but the researchers had no time to create a uniform plan. From Maryland to Nebraska, a dozen research teams fanned out across the United States and Canada to glean whatever facts they could to fill in the many gaps in our understanding of the sliver of the agro-ecosystem where corn, milkweeds, and monarchs meet.

The Biotechnology Industry Organization was eager to dispel public fears about the lethal effects of Bt corn on butterflies—and to head off any restrictive new EPA rules. BIO asked the scientists involved in the summerlong research effort to pull their findings together as quickly as possible in the fall. Even though most of the scientists would have had a chance to analyze only a small portion of the data collected by then, the biotech industry set a meeting of the full Agricultural Biotechnology Stewardship Working Group (as the research venture was called) for the beginning of November.

At seven-thirty in the morning on November 2, 1999, more than a hundred scientists from industry, academia, the EPA, and the USDA packed into a meeting room in a hotel within earshot of Chicago's O'Hare Airport, to take part in the industry-sponsored monarch butterfly research symposium. "What the group hears today is very likely to have an impact on future growing seasons for Bt corn," Janet Andersen began. EPA scientists, she explained, had not tried to feed the toxic substance produced in Bt corn to monarch caterpillars or any other butterfly species. Instead, they had made the assumption that the Bt toxin could be poisonous to all butterflies and moths. They then asked themselves: What are the chances that any butterflies and moths will ingest Bt corn?

"EPA concluded that it was highly unlikely that endangered, threatened, or even highly revered species such as the monarch would have a significant exposure to Bt corn," Andersen told the hushed group of industry and university scientists. "New data on migration pathways, feeding patterns, or any of the other information we may hear today may lead us to alter that conclusion."

Orley R. "Chip" Taylor, chairman of the department of entomology at the University of Kansas in Lawrence, opened the scientific portion of the meeting by making an impassioned plea to the industry and to government regulators to protect monarch butterflies. He described the annual fall monarch butterfly migration as one of the most magnificent and mysterious events in the natural world. Taylor runs a large tagging program, called Monarch Watch, that enlists the help of thousands of schoolchildren around the country each fall to track the monarch butterfly migration.

"This is the most spectacular migratory insect phenomenon known," he said. "These butterflies go to Mexico, where they overwinter on a few mountaintops about a hundred miles west of Mexico City. The phenomenon is truly marvelous. There are somewhere around sixty to a hundred million butterflies that overwinter there each winter."

Taylor laid out some of the questions that biologists would have to answer before they could determine the true extent of the threat Bt corn poses to monarch butterflies. The scientists enlisted by the biotech industry had just begun to tackle a few of those questions, he said. "We have to confront the reality that we can't really make an assessment until we know some basic things about the Bt effect and about the monarch butterfly," he said. "We have to know, for example, the persistence of the toxin. We don't know if we're dealing with a momentary event out there or something that lasts for thirty days. We don't know the concentrations that kill or reduce fitness in the monarch. We can identify the kill, but we are going to have a hard time looking at the reduced fitness of the butterflies from partial exposure to Bt pollen.

"More seriously, we really don't understand the annual and seasonal dynamics of the monarch butterfly very well," Taylor told the symposium participants. "This population fluctuates a great deal from year to year, and we're not really totally competent to understand why."

For the next several hours, more than a dozen scientists presented summaries of the field studies they had performed over the summer, from counting milkweed plants on roadsides to calculating how many Bt pollen grains it took to weaken or kill a monarch caterpillar.

Robert Hartzler, an agronomy professor at Iowa State University, set out during June and July with a USDA colleague, Doug Buhler, to identify where the greatest concentration of milkweed plants were to be found in Iowa. They found common milkweed growing on about 70 percent of roadsides and, unexpectedly, within about half the corn and soybean fields they surveyed. There were more milkweed plants per square meter on roadsides than in corn and soybean fields, but since corn and soybean fields make up more than three-quarters of Iowa's land mass, cornfields may be a surprisingly important resource for monarch butterflies.

Mark Sears, chairman of the department of environmental biology at the University of Guelph's Ontario Agriculture College in Canada, had spent the last two weeks of July measuring the amount of pollen that flows from cornfields onto nearby plants. By the day of the symposium, he had completed the counts for only one cornfield, in Woodstock, Ontario. Based on his preliminary data, Sears concluded that monarch caterpillars are likely to eat Bt corn pollen on milkweed plants growing within one to two meters of the edge of a cornfield.

Galen Dively of the entomology department at the University of Maryland in College Park measured how much corn pollen was deposited on milkweed plants at various distances from Maryland cornfields. He and his assistants removed more than a thousand leaves from 572 milkweed plants growing in or around cornfields in six locations in the state. During the course of their fieldwork, his research team found no monarch caterpillars feeding on any of the milkweed leaves. This observation led Dively to conclude that larvae feeding did not coincide with the time when corn pollen was being shed. At least in the summer of 1999 and in the Maryland locales he studied, monarch caterpillars appeared to face little threat from Bt corn pollen.

Taylor described his attempt to estimate the threat that monarch butterflies faced from the millions of acres of Bt corn grown in America over the summer. He first determined the overall monarch breeding habitat for North America and then estimated the amount of prime monarch butterfly habitat that lies within the farm belt. Taking into account estimates of the number of eggs laid per acre, the amount of Bt corn in cultivation, and the acreage where milkweed plants might be growing within the corn pollen shadow, he tried to calculate how much potential "add-on mortality" the monarch population might suffer from Bt corn.

He predicted a potential loss of 7 percent of the fall migratory monarch population due to the toxicity of pollen from 25 million acres of Bt corn, but added that "if we increase crop production to forty million acres, we could lose eleven percent." Taylor concluded that "the potential impact is significant; however, the realized impact will be much less."

After the scientists finished their presentations, one academic researcher in the audience asked Janet Andersen how many species of Lepidoptera larvae the EPA had tested against Bt corn pollen.

"Our scientists know that Bt proteins are capable of being toxic to lepidopterans, and we made the assumption that it would be toxic to all lepidopterans, okay? We can make that kind of assumption. We didn't test lepidopterans," she responded. "Now, if we have made a mistake, then through the process we can go back and make modifications."

The biotech industry's $100,000 research project had shed some light on the agro-ecosystem where monarch butterflies coexist with cultivated cornfields. The scientists had made a valiant start at probing the questions surrounding the effect of Bt corn pollen on one species there. As the researchers themselves were quick to acknowledge, however, their findings raised scores of new questions about the complex interaction between monarchs, milkweeds, and Bt corn.

"This all should have been settled before EPA allowed millions of acres of Bt corn to be planted," one scientist in the audience lamented.

None of the scientists participating in the Chicago symposium claimed to have a definitive answer to the question of whether Bt corn poses a real threat to monarch butterflies in the wild. Richard Hellmich characterized the summer's studies as a good "first step." Yet the biotechnology industry used the gathering to launch a sneak public relations coup.

Midway through the afternoon, it came to light that the Biotechnology Industry Organization, on behalf of the Agricultural Biotechnology Stewardship Working Group, had faxed a press statement to newspapers across the country claiming that the scientific panel—still under way—had already concluded that Bt corn "poses negligible harm to the Monarch butterfly population." The media press release went on to say that new field research "dispels doubts raised last spring about the safety of the Monarch population."

The scientists at the symposium had drawn no such conclusion. In fact, those of us attending the Chicago meeting learned about the media advisory only when Carol Kaesuk Yoon, a reporter from *The New York Times,* stood up and said her editor had received the release. A heated discussion ensued. Many of the researchers were dismayed that the biotechnology industry had put its own "political spin" on what was supposed to be a purely scientific meeting.

Taylor insisted that the scientists had "absolutely not" come to the conclusion that monarchs are safe from Bt corn. Another researcher agreed with Taylor. While the risks to monarchs were probably "not that great," he said, scientists would not be able to know for certain until all of the missing pieces of research were assembled: "We're a long ways from determining what the exact risk is."

After attempting to dispense with the possible threat that Bt corn pollen posed for monarch butterflies, the Biotechnology Industry Organization turned to its next major public relations battle: the looming World Trade Organization meeting. Preparations for these international trade negotiations had been in the works for nearly a year. In January 1999 BIO had joined with the American Farm Bureau Federation and other key agribusiness trade groups to form the Seattle Round Agriculture Committee (SRAC), to represent U.S. agribusiness interests at the WTO's Third Ministerial Meeting. By November the coalition had gathered nearly ninety powerful U.S. food and farm associations, agribusiness companies, and state departments of agriculture to speak with a united voice on the direction that American food and agriculture policies should take. The SRAC corporations, including Archer Daniels Midland, Campbell Soup, Cargill, ConAgra, Monsanto, and Philip Morris, wanted a further liberalization of agricultural trade.

A key goal of the SRAC was to use the Third Ministerial Meeting to require the European Union to approve genetically engineered crops. In the weeks leading up to the meeting, the mighty agriculture coalition envisaged a prestigious event for international trade in Seattle, where the full weight of global industries would be on display.

GLOBAL FOOD FIGHT

W hat took place at the World Trade Organization meeting was not an event scripted by corporate power brokers and trade ministers but something much larger, of historic magnitude. On November 30, 1999, consumers demanding the right to know what kind of food they were eating, workers seeking to reclaim their livelihoods from child laborers overseas, and environmentalists defending wildlife converged in Seattle in a spontaneous outpouring of protest. Ignited by a fringe group of anarchists, their protests exploded on the downtown streets with an intensity not seen by Americans in nearly thirty years.

The weather was chilly and overcast, but throngs of protesters, many in theatrical costumes, filled the streets, lending a carnival atmosphere to the city. Approaching the downtown on foot, I encountered demonstrators of all ages, some chanting and beating drums, others carrying placards, pushing floats, or dangling marionettes, creeping along through clogged thoroughfares. Scores of young people wearing handcrafted orange-and-black monarch butterfly wings were seeking to halt the planting of biotech corn. "Hey hey, ho ho, GMOs have got to go!" they chanted. Dozens of college-age students wore jade-green cardboard turtle shells on their backs to protest destruction of the marine habitat. A large contingent of middle-age workers carried signs reading "Globalize Workers Rights" and "Fair Trade, Not Free Trade." Hundreds of twentysomething men and women in street clothes chanted and brandished signs saying "Resist Corporate Tyranny," while others shouted, "No globalization without representation!" Monsanto, frequently transliterated to "Mutanto" on placards, was one of the chief global corporations the protesters targeted.

Thousands of demonstrators choked the streets leading to the Sheraton Hotel, where delegates and reporters were supposed to register for the WTO meeting. Inching my way through the crowd, I was stopped at an intersection near the Sheraton by a line of protesters linked arm in arm. I asked them to let me through, but none would break ranks.

A phalanx of police garbed in black riot gear, helmets, and face shields and wielding an impressive array of weaponry stood immediately behind the human chain of demonstrators. Assuming that the police had cleared a corridor at one of the entrances to the meeting, I asked an officer where I should go to get access. He stared straight ahead, impassively, refusing to acknowledge me.

Near Fifth Avenue and Pike Street, I again tried to find an opening in the crowd and was again met with resistance. The protesters had formed tightly knit human chains across all the streets encircling the Sheraton Hotel, the Convention and Trade Center, and other main venues for the WTO summit. At around three o'clock, near the intersection of Pike Street and Fourth Avenue, the highly charged mood of the crowd abruptly turned ominous. A crush of demonstrators ran down the street screaming. Simultaneously, a loud percussive boom rocked the air, coursing through my body like an electric shock. A blanket of thick white fog unfurled along the pavement as police fired canisters of tear gas into the crowd.

Pandemonium erupted, and I joined the throngs of people streaming downhill, away from the melee. A tall, slender college-age boy ran up beside me shouting, "I've been shot. Our government is shooting at us!" Grimacing, he pulled up his pant leg and rubbed his calf. Another young man, apparently one of his companions, stopped up short behind us. "Rubber bullets!" he shouted breathlessly. "The police are shooting rubber bullets!" A third member of their group appeared, cradling a marble-size plastic pellet in his cupped hand. The police were apparently firing plastic bullets into the crowd.

"I can't believe the government is shooting at us!" the stunned

youth mumbled in disbelief. Then he rolled down his jeans and turned
to rejoin his friends, who were already scrambling back up the hill
into the combat zone.

Throughout the afternoon, the chaos intensified as the police used
pepper spray and tear gas to try to disperse the thousands of protesters jamming the streets. The vast majority of demonstrators, including
an estimated forty thousand labor union members, environmentalists, and consumer activists, had staged peaceful marches and rallies
throughout the day. A small group of rowdy anarchists, however, had
rampaged through the commercial district. Dressed in black hooded
sweatshirts, they and other violent marauders hurled stones and newspaper boxes through the glass window fronts of retail stores.

At five-thirty P.M. Mayor Paul Schell declared a state of emergency
in downtown Seattle. The governor then called in two hundred members of the National Guard. As skirmishes between protesters and police persisted, Schell ordered a curfew from seven that evening until
seven-thirty the following morning.

A local cable TV station broadcast snippets of the WTO meeting.
U.S. trade representative Charlene Barshefsky offered profuse apolo-
gies to the foreign delegates, many of whom had traveled great distances at tremendous expense, only to be mauled or blocked from
entering the trade talks.

"This has not stopped our work," WTO Director-General Mike
Moore insisted hollowly, as mayhem raged on the streets outside. In a
humiliation for the Clinton administration—and a victory for protesters who had vowed to shut down the WTO—neither Secretary of
State Madeleine Albright nor UN secretary general Kofi Annan was
able to reach the formal opening session in the Paramount Theater.

The World Trade Organization, which had 135 members in 1999, was
created in 1995 with the goal of spreading the benefits of free trade
around the globe. The Geneva-based trade group sets rules aimed at
breaking down barriers to the flow of products between nations and
trading blocs. A WTO appellate body mediates trade disputes between member nations and has the power to levy fines against countries that refuse to abide by its rulings.

The activists squatting outside the Seattle meeting viewed the WTO as a vehicle to help global corporations consolidate their business monopolies and dominance over the lives of ordinary citizens. The demonstrators had diffuse targets: fast-food restaurant chains and retail stores that homogenize food and taste; nations that allow businesses to employ child labor; corporations and nations that pollute the air and water or endanger vulnerable species; and banks and international organizations that finance development projects that hasten the destruction of ecosystems and strip indigenous peoples of their natural resources.

Domestic and international consumer groups opposed to America's stance on genetically engineered foods made up a large proportion of the demonstrators. Many directed their anger at Monsanto, and by extension at the U.S. government, for attempting to force those foods into countries where the public did not want to eat them.

Demonstrators from numerous environmental groups were protesting U.S. efforts to derail global negotiations for a Biosafety Protocol. The effort to create a Biosafety Protocol had grown out of the Convention on Biological Diversity signed in Rio de Janeiro in 1992. Some drafters of the convention were worried that genetically engineered fish, animals, and food plants would threaten the environment, especially in developing nations rich in biological diversity. The concern is greatest for regions of the world that are "centers of diversity" for major food crops—places where the crop has been grown for thousands of years and has many wild relatives. Mexico and Peru are centers of diversity for corn. The Andean regions of Peru, Ecuador, and Bolivia are centers of diversity for potatoes. China is a center of diversity for soybeans, sugar cane, oranges, tea, and millet, while North America is a center of diversity for sunflowers, berries, black walnuts, and pecans. Crop diversity has proven essential to the survival of agriculture. When farmers experience a major disease or pest outbreak in a crop such as corn or potatoes, plant breeders go to centers of diversity for that crop to search for pest-resistance traits in the plant's wild relatives. If genetically engineered plants are released near wild relatives, the altered plants might threaten the gene pool of native plants.

Formal international negotiations on the Biosafety Protocol began in July 1996, with the goal of setting tight controls on international movement and trade in genetically modified organisms, referred to as "living modified organisms," that might harm biodiversity. The European Union and most of the other global signatories to the Convention on Biological Diversity wanted to adopt the precautionary principle as a legitimate ground for countries to reject the importation of genetically engineered crops or products. An overwhelming majority of countries had been poised to adopt the Biosafety Protocol in Cartagena, Colombia, in February 1999, when the United States led five other grain-exporting nations—Argentina, Australia, Canada, Chile, and Uruguay—in torpedoing the negotiations. Known as the Miami Group for the city where the alliance was formed, the six-nation bloc insisted that shipments of living modified organisms that were commodities for use in food or feed should be exempted from the protocol.

For all their disparate causes, one sentiment unified the environmental activists and labor union members who had flocked to Seattle: a deep mistrust of global corporations. Ralph Nader decried the WTO system, charging that it was based not on the health and economic well-being of people but "on the enhancement of the power and wealth of the world's largest corporations and financial institutions."

The WTO allows trade bureaucrats in Geneva to make a host of policy decisions that were once handled exclusively by local and national governments, Nader charged. "These bureaucrats, for example, are now empowered to dictate whether people in California can prevent the destruction of their last virgin forests or determine if carcinogenic pesticides can be banned from their food, or whether the European countries have the right to ban the use of dangerous biotech materials in their food."

Nader argued that the U.S. Congress, like the legislative bodies of other WTO-member nations, had "ceded much of its capacity to independently advance health and safety standards that protect citizens." The WTO, he concluded, is a "race to the bottom" by countries, not only in the deterioration of health and environmental safeguards but in democracy itself.

On December 1, the second morning, I managed to slip onto a bus chartered by the WTO to ferry delegates to and from the meeting, by putting my congressional press pass around my neck and striking up a conversation with another delegate as he was boarding. The driver snaked through several police barricades on the short trip to the Sheraton Hotel, where I was finally able to pick up my WTO credentials. For the next three days, the plastic identification badge was virtually all that separated me from jail. Invoking civil emergency powers, Mayor Schell had designated a wide swath around the convention center and major downtown hotels as a "no-protest zone." A half dozen National Guardsmen stood sentinel at each intersection within the fifty-block zone, stopping pedestrians and demanding to see identification. Anyone without a WTO badge or proof of employment or residence within the zone was required to move away immediately. Those who refused were arrested. The police arrested about four hundred protesters that day.

The official WTO meeting consisted of dozens of concurrent negotiating sessions, which were carried out mostly in secret. Periodically, trade ministers would issue communiqués or stage spontaneous press conferences, especially when things were not going their way. A second, much more open meeting was made up of simultaneous gatherings and press briefings called by nongovernmental organizations (NGOs), whose groups all had a stake in the outcome of the official talks. Trade ministers emerging from behind closed doors tended to speak elliptically, if at all. NGOs, eager to get their message into print, were straightforward and vocal.

The goal of the Seattle "Millennium Round," as the Third Ministerial Meeting was called, was to set an agenda, or framework, for international trade liberalization talks for the coming three years. The Clinton administration's major objective was to improve the U.S. position in agriculture trade. The challenge of finding an agenda agreeable to all parties to the WTO, Agriculture Secretary Dan Glickman told reporters on December 1, grew out of the fact that each of the world's distinct blocs has its own unique agricultural policies and interests. The fifteen EU nations tend to have small farms with high per-unit production of crops. Japan has more than three million farmers and exports few agricultural products. Canada and several Latin American nations are big grain exporters and oppose export subsidies. The developing world generally lacks the skills and technical capability to operate in the global market economy. One of the main objectives of the U.S. negotiating team was to use the WTO rules as a way to speed approvals of genetically modified foods in Europe and make it harder for countries to reject those imports.

During a speech at the Port of Seattle, President Clinton addressed global fears about the safety of America's genetically modified foods. "We have an orderly disciplined system here for evaluating the safety of not only our food but our medicine," the president said. "And we ask all of our trading partners to deal with us in a straightforward manner about this. But everybody has to understand we have nothing to hide, and we are eating this food too."

The biggest player in the NGO field was the Seattle Round Agriculture Committee (SRAC), representing ninety U.S. food and agribusiness trade groups. This influential industry group had access to the U.S. trade representatives and top government officials. SRAC wanted the next round of trade liberalization talks to produce rules that would require the EU to speed up the stalled approval process for genetically modified foods. "WTO rules must ensure that approval processes for biotech products are transparent, predictable, timely and science-based," the SRAC stated.

"Countries have the right under WTO rules to protect themselves from legitimate food safety and environmental risks. They can, when necessary, restrict the importation of hazardous products. However, they also have the obligation to demonstrate that such regulations are not arbitrary, politically motivated or discriminatory, but are based on sound scientific evidence," the SRAC continued. "The new round of negotiations should reinforce this principle with respect to the products of biotechnology."

Missouri senator Kit Bond was in step with his state's agribusiness and biotech industries. At a December 1 press briefing organized by the SRAC, Bond released a letter signed by three hundred scientists from industry, academia, and government attesting to the benefits of agricultural biotechnology. "In the near future, an ever-increasing number and variety of crops with traits beneficial to consumers will reach the market," the scientists said. "Such traits will include improved nutritional values, healthier oils, increased vitamin content, better flavor and longer shelf life. In developing countries, biotechnological advances will provide means to overcome vitamin deficiencies, to supply vaccines for killer diseases like cholera and malaria, to increase production and protect fragile natural resources, and to grow crops under normally unfavorable conditions."

When asked, at the same briefing, what progress could be expected on getting the Europeans to accept biotech crops, Ambassador Peter Scher acknowledged that that was one of the toughest issues facing U.S. negotiators. The U.S. trade representative hinted that any text to arise from the Millennium Round talks might fall far short of the outcome American growers wanted.

A major point of contention between the United States and the European Union centered on two tortuous terms: *multifunctionality* and *precautionary principle*. The Clinton administration had initially sought to keep the phrases out of any agreed-upon negotiating text, but the Europeans insisted on their inclusion.

Multifunctionality, as European delegates explained the concept, is an approach to farming that takes into account more than simply the production of livestock and crops. Europeans view farming as critical to maintaining the culture, history, and ecology of the countryside. They support farmers for their role in preserving a cherished way of life as well as producing food. Seen through the lens of multifunctionality, the large subsides that European governments pay farmers to keep their rural landscapes intact are not, strictly speaking, "export subsidies" but rather a form of social preservation.

American agribusiness takes a singular, hard-nosed approach to the purpose of farming, namely, to produce an abundance of food, feed, and fiber as cheaply and efficiently as possible. U.S. agricultural trade groups dismissed the concept of multifunctionality as little more than a canard to cloak the estimated $50 billion in farm subsidies that the fifteen nations of the EU pay their farmers each year.

Dean Kleckner, president of the American Farm Bureau Federation, took a dim view of multifunctionality. "A country may want to pay a farmer for keeping his roadside mowed, for building stone fences or for rural employment purposes," he told the WTO. "If countries want to pay us for being cute and quaint, that's fine—as long as the payments are reasonable and do not distort trade. I say to those nations that pursue such practices: keep as many inefficient farmers around as you want, just don't expect to dump their products into world markets at unfair prices."

The precautionary principle, as European delegates described it, is a scientific approach to the regulation of potentially dangerous technologies. The principle comes into play when a new technology presents society with a serious or irreversible threat about which scientific evidence is lacking. In such instances—notably the introduction of genetically engineered crops into the environment and food chain—the precautionary principle requires proponents of the new technology to prove its safety. Rather than placing the onus on consumers and regulatory agencies to demonstrate that a safety problem exists, the precautionary principle shifts the burden of proof to the companies that created the technology.

The precautionary principle is a radical departure from traditional risk assessment approaches. U.S. officials have described risk assessment as a tool—combining the art and science of estimation—that allows a new technology go forward before firm data on its outcome are available. Risk assessors try to calculate—and balance—the benefits of a new technology, or chemical, or activity against potential harm to public health and the environment. Risk-benefit equations often factor in economic and social costs. Risk experts also sometimes set an "acceptable" level of negative outcomes—one cancer death per million people exposed to a toxic agent, for instance—in exchange for an assumed positive societal benefit from use of the agent.

Many U.S. environmentalists and organic farmers have argued that too little information is available on the effects of gene-altered foods on humans and the environment to develop any true risk assessment models.

In Europe agitation for use of the precautionary principle with genetically engineered crops came not from bureaucrats but from consumers and environmental groups like Greenpeace. Benedikt Haerlin, international coordinator of the Greenpeace Genetic Engineering Campaign, had provided his views on the precautionary principle at a science meeting in London two weeks before the WTO: "Present scientific knowledge about the functionality of DNA, and even more, about the dynamic interaction between organisms are completely insufficient to assess the possible impact of the release of such new life forms into the environment. The combination of massive uncertainties about possible detrimental effects of GMOs in the environment and the global long-term potential of such possible effects, leads us to demand the strictest possible application of the precautionary principle."

Greenpeace is not advocating a risk-free world, Haerlin insisted, "but when it comes to risks taken on the public, on the environment, and on future generations, risks taken on people who have not been informed and not asked for their opinions, we take a precautionary approach."

European critics prodded their health and safety regulators to demand that safety tests be performed on genetically engineered foods by independent scientists. Those who are closest to the development of a new technology are not always the best, the least biased, or the most cautious advisers on the risks of the technology, Haerlin noted. "Neither scientists nor industry appear to be ideal arbitrators of the desirability and the risk-benefit analysis" of the new technology.

Most U.S. food and agribusiness leaders view the precautionary principle as antiscience. Some industry representatives accused European politicians of pandering to public hysteria and irrational fears. Others called it "voodoo science." Use of the precautionary principle in the approval process for genetically modified crops would lead to paralysis for years in Europe, some biotech industry officials predicted. Many industry analysts have warned it would turn back the clock on scientific discoveries and technological innovation. "Make no mistake, the precautionary principle is a wolf in sheep's clothing," one food industry official later told me.

As I left Senator Bond's agriculture briefing at the Madison Hotel to walk the few blocks to the convention center, a group of people milling just outside the invisible police perimeter were shouting, "Assert your rights, the right to assemble!" Six National Guardsmen dressed in full riot gear stood on the sidewalk. One asked to see my ID. "Okay, go on," he said, as I pulled aside my coat lapel, revealing the WTO badge hanging from my neck.

"This is what democracy looks like! This is what democracy looks like!" chanted another cluster of youth gathered on a street bordering the no-protest zone. A group of riot police approached them, night sticks erect.

Another NGO, the U.S. Consumer's Choice Council, sponsored a session on genetically engineered food at which Mae-Wan Ho, a geneticist and biophysicist in the UK, delivered an open letter to the WTO delegates. The World Scientists' Statement, signed by 140 scientists from twenty-seven countries, called for a five-year moratorium

on genetically modified crops. Genetic engineering, as currently practiced, Ho told reporters, is inherently hazardous because it is based on
a flawed paradigm that is "out of date and in conflict with scientific
findings."

Ho began her thirty-year career in genetics as a postdoctoral fellow
in biochemical genetics at the University of California in San Diego
and has served for years as a lecturer in genetics at the Open University in London. In 1994 she began to have grave misgivings about the
use of genetic engineering to alter food plants. "The existing technologies are crude, unreliable, uncontrollable and unpredictable," she
said.

Under the old model of genetics, each organism is seen as "a collec-
tion of traits, each tied to specific genes which do not, by and large, interact with one another, nor with the environment, and these genes are
passed on unchanged to the next generation—except for very rare
random mutation," Ho said. "If this were true, then genetic engineering would be as precise and effective as is claimed. Unfortunately,
scientific findings within the past twenty years reveal an immense
amount of cross-talk between genes which function in complex networks. Genes are nothing if not sensitive, dynamic, and responsive to
other genes, to the cell or organism in which they find themselves, and
to the external environment."

In addition to relying on outdated ideas about how genes function,
genetic engineering is dangerous because it has been molded by a
handful of corporations acting from the profit motive, Ho charged.

The 140 signatories to the petition included forty U.S. scientists.
They described several recent scientific studies that cast doubts on the
theoretical underpinnings of genetically engineered food. One disturbing study showed that DNA is not broken down as rapidly in the gut
as research in the 1970s and 1980s had suggested. In January 1999
New Scientist magazine reported on a computer-controlled model of a
stomach and intestines, designed by Dutch researchers, that mimics
the process of human food digestion. Robert Havenaar, who created
the simulated gut, found that DNA lingers in the intestine long
enough for genes to transfer to bacteria in the stomach of the person
eating gene-altered food. Havenaar found that DNA from genetically
modified bacteria had a half-life of six minutes in the large intestine—

enough time, potentially, for human bacteria to become transformed by the foreign DNA.

Ho charged that the "promises to genetic engineer crops to fix nitrogen, resist drought, improve yield and to 'feed the world' have been around for at least thirty years. Such promises have built up a multibillion-dollar industry now controlled by a mere handful of corporate giants.

"The miracle crops have not materialized," she said, rebutting Senator Bond's claim that biotech companies are on the verge of introducing crops with traits that deliver a wide array of benefits to consumers. So far, two simple characteristics continue to account for virtually all the genetically modified crops in the world. More than 70 percent are tolerant to broad-spectrum herbicides, while nearly all of the rest are engineered to kill insect pests.

Back at the convention center, the most intriguing diplomatic maneuvering between the United States and the European Union centered on the creation of a WTO working group on biotechnology. The U.S. and Canadian negotiators tried to persuade the Europeans to agree to such a working group within a new round of WTO trade negotiations as a way circumvent the rules on trade in genetically modified organisms being negotiated under the Biosafety Protocol. The existence of a WTO working group on biotechnology, analysts believed, would allow countries that export genetically modified food products to call for a suspension of the Biosafety Protocol until the WTO had time to clarify its own rules. Ongoing parallel talks on biotech food safety under two separate international agreements—one, the WTO, governing trade, and the other, the Biosafety Protocol, governing the environment—could have led to a protracted dispute over which agreement took precedence over the movement of living modified organisms among countries.

The European Commission, the executive branch of the EU, created a stir when it reportedly acquiesced to the U.S. and Canadian idea. The environment ministers from Belgium, Denmark, France, Italy, and the UK swiftly rejected the working group, insisting that the proper forum for deciding on trade rules involving genetically modified organisms was the talks on the Biosafety Protocol. The ministers issued a formal statement disavowing a WTO working group on biotechnology, arguing that by potentially subordinating the environmental agreement to a trade agreement, it would "run directly counter" to the key objective of the Biosafety Protocol.

While delegates from the world's major trading blocs bickered over how to evaluate the safety of genetically engineered foods, 130 antibiotechnology activists from twenty countries were of one mind on the subject. Activists from Australia, Austria, Brazil, Canada, Ecuador, France, Germany, India, Japan, Kenya, Korea, Malaysia, Nicaragua, Paraguay, South Africa, Singapore, Sweden, Switzerland, the United Kingdom, and the United States agreed to the need for a global ban on all genetically engineered processes, foods, crops, and animals and called for full labeling of all bioengineered foods.

In addition, the activists tackled the subject of "biopiracy," wherein biotech companies and researchers obtain seeds, plants, tissues, blood samples, and other genetic materials from farmers and peasants in developing nations and then use them to make patented genetically modified food plants and medicines. The environmental NGOs called for criminalizing the "acquisition of indigenous genes." They also pressed the governments of the world to establish "strict corporate liability for all economic loss and personal injury resulting from genetically engineered crops and food."

On Friday night, December 3, 1999, the WTO talks broke down irreparably—to great jubilation and cheers from the protesters outside on the streets. "The issues that remained were highly complicated and, we believe, could not be overcome rapidly," U.S. Trade Representative Charlene Barshefsky told reporters.

Delegates from developing nations claimed they had been excluded from critical meetings and negotiating sessions. Smaller and poorer nations felt they were placed at a disadvantage in the talks because they could afford to send only a few delegates to Seattle, whereas the United States, the European Union, and Japan each had more than seventy-five negotiators on site. Trade ministers from the Organization of African Unity and some Latin American and Caribbean nations complained there was a lack of transparency at the meeting. And the gulf between the United States and the European Union over biotech crops and export subsidies proved too wide to bridge.

The ill will fomented inside the formal meeting halls of the WTO throughout the week contrasted notably with the common bonds forged outside on the street and in jail cells. Traditionally, the goals of environmentalists, seeking to protect endangered species and habitats from development and communities from pollution, were depicted as at odds with the goals of labor organizations, which primarily wanted to preserve industry and their livelihoods. Labor unions saw global corporations as transplanting high-paying factory jobs from the United States to countries with a cheap labor pool that often included children. In Seattle many labor activists and environmentalists embraced the common goals of protecting human health, the environment, and jobs from corporate policies they felt often paid too little attention to all three. The new alliance of environmental groups and labor unions was perhaps the most remarkable outcome of the week.

Another important outcome of the week of protests was an increase in awareness of genetically engineered foods among activists in Seattle as well as among U.S. television viewers. Consumers seeking labeling or a moratorium on planting made up just one segment of the protesters, but the official U.S. opposition to segregation and labeling—despite widespread popular appeals—emerged as a clear example of how global corporations can use the power of governments to impose their own products and policies on unwilling consumers around the world.

When representatives of more than 130 nations traveled to Montreal in mid-January 2000 to try again to reach agreement on a Biosafety Protocol, the world community displayed a much firmer resolve to regulate commerce in genetically modified products. In the six weeks between the breakdown of the WTO talks and the Montreal meeting, public opinion in many regions of the world had coalesced against the new foods.

During the months leading up to the WTO talks, some international consumer groups in the Asia-Pacific region had begun to voice worries about the safety of genetically modified foods—concerns that hardened after the Seattle demonstrations. In October 1999 Yoko

Tomiyama, chairperson of the Consumers Union of Japan, posted an open letter to American Farmers and Agribusiness. The letter was a plea for North American growers to plant and segregate enough non-engineered corn, soy, cotton, potato, and canola in the 2000 growing season to meet the demands of Japanese consumers. "While some would have you believe that our opposition to GMOs is a trade barrier," Tomiyama wrote, "this is untrue. Japanese consumers are seriously concerned about the potential health and environmental hazards of GMOs. We believe that there is solid scientific basis for our concern."

Japanese consumers asked American farmers to ship them grain with less than 0.1 percent GMO content. In August 1999 the Japanese government decided to require labeling of genetically modified crops and foods containing genetically modified ingredients. The labeling program was set to take effect in April 2001. Japanese companies were mindful of consumers' wishes, however, and started taking the ingredients out of their foods well before the labeling deadline was imposed. Kirin and Sapporoto Breweries of Japan announced they were removing genetically modified corn from their beer. Kanot Soybean Wholesale of Japan said it would purchase only nonengineered soybeans. The Japan Tofu Association pledged to use only nonengineered soybeans in its products.

Korean consumers too were eager to secure nonengineered sources of corn and soybeans. The Korean government approved mandatory labeling of genetically modified corn, soybeans, and bean sprouts, effective in March 2001.

In August 1999, responding to public demands, the Australia New Zealand Food Authority (ANZFA) decided to order mandatory labels on genetically modified foods. The following year, in August 2000 ANZFA health ministers adopted a strict zero-tolerance standard for labeling, to take effect in mid-2001. The standard was passed over the strenuous objections of the food industry. Many consumer groups in Australia charged that the labeling scheme did not go far enough, however, since it exempted foods served at restaurants. Labeling laws in Europe apply to genetically modified foods sold in cafés and restaurants as well as grocery stores and markets.

The labeling initiatives in Japan, Korea, Australia, and New Zealand, coupled with the fervent protests in Seattle, made it clear to delegates at the Biosafety Protocol negotiations in Montreal that support for genetically engineered foods was eroding around the world. The Miami Group tried again to subvert the Biosafety Protocol. Canada argued on behalf of the United States (which had never ratified the Convention on Biological Diversity and hence could not participate in the talks) that living modified organisms destined for use in food and feed should be exempted from the agreement. The Miami Group also argued that countries should not be able to use the precautionary principle in deciding on the environmental and human health effects of genetically modified organisms. This time the 132 countries in favor of the precautionary principle and regulation of movements of the new foods prevailed over the five voting members of the Miami Group. On January 29, 2000, representatives of more than 130 countries adopted a Protocol on Biosafety under the UN Convention on Biological Diversity, leaving the United States virtually isolated in its position on genetically engineered foods.

The fallout from Seattle had immediate repercussions for the U.S. government closer to home as well, as increasing numbers of American consumers, organic producers, and conventional farmers began to agitate for biotech crop regulation and food labeling.

ELEVEN

SEEDS OF DISPUTE

T en days after the WTO protests, Oakland, California, became the site of another demonstration against genetically modified food. On December 13, 1999, more than a thousand people turned out for a noon rally outside the city's federal building to protest the FDA's failure to require safety testing and labels. Inside, FDA deputy commissioner Sharon Smith Holston faced a spillover crowd of angry consumers and organic farmers as well as representatives from biotechnology companies and university laboratories. The meeting was the last of three public hearings that the FDA conducted that fall. More than a thousand people had attended the first two meetings, which were held in Chicago on November 18 and in Washington, D.C., on November 30. The meetings were intended to explain how the government oversees genetically modified foods and to give members of the public a chance to express their views to the FDA on how the foods should be regulated.

Susanne Huttner, director of biotechnology research at the University of California at Berkeley, was one of the first speakers the FDA had invited to address the crowd. She argued in support of the FDA's voluntary system for reviews of biotech food products. The FDA does not routinely subject foods from new plant varieties to specific safety tests, she explained. Instead, companies creating a genetically engineered food are supposed to use a "decision tree" to decide whether their new product is safe or presents possible allergenic risks.

"Any risks from genetically engineered organisms in foods can be managed using the existing system, without adding unnecessary cost burdens to small business or poor consumers," Huttner said. "It's working very well the way it is right now, and you have a good deal of information that's available and will continue to be made available by companies."

Huttner said she had consulted with the Centers for Disease Control in Atlanta and could not find evidence of anybody being made sick from eating a genetically engineered food. In fact, the government has not set up a registry to track possible health effects related to the ingestion of foods made with genetically modified corn, potatoes, soybeans, or canola.

Because the government had not required biotechnology companies to conduct long-term animal feeding studies or human clinical trials on their gene-altered corn and soybeans, Huttner relied on the ongoing exposure of the U.S. population to provide evidence that the foods are safe. Since the 1996 harvest, most men, women, and children in America are presumed to have been exposed to genetically modified corn and soybeans on a regular basis with no apparent ill health effects. "Literally millions of people have eaten the products of genetic engineering over the last decade," Huttner asserted, "and there isn't a single instance of someone being made sick by these products."

During the day-long meeting, the FDA allowed eighty members of the public to speak for two minutes each. Ignacio Chapela, an assistant professor of ecology at the University of California at Berkeley, warned that scientists currently have only a "primitive understanding of the consequences of the recent manipulations of plant, animal, and microbial life in the new agricultural biotechnologies. We know enough only to know that the new genetic engineering methods have definite and potentially enormous risks to the environment." Chapela noted that many of his colleagues have been forced to enter into research agreements with life sciences companies in order to obtain "access to funds that they consider necessary for their professional survival."

California is home to many of the nation's six thousand certified organic growers, and a number of them used their allotted time to talk about the impact of genetic engineering on their livelihoods.

Organic foods from pesticide-free lettuce to free-range chickens have been growing in popularity among consumers in the United States and abroad. But organic farmers said they face a serious threat from "biological pollution"—the contamination of their crops by pollen from nearby gene-altered varieties. Organic growers told the FDA that they can control their soil, their inputs, and their methods to meet the demands of their buyers for pure foods, but they cannot control bees, butterflies, and wind, all of which can deposit pollen from genetically engineered plants on their varieties.

Jo Ann Baumgartner, half owner of Neptune Farms and an organic farmer on the California central coast for fifteen years, told the FDA officials that "drift from unregulated GE crops is causing genetic pollution. Organic and conventional non-GE farmers should not have to suffer to accommodate these Frankenstein foods." Gene-altered foods should be labeled not only on the grocery store shelves, she argued, but in the fields, so that nonengineered-food growers would know what their neighbors were planting and have the opportunity to decide whether to sow their own seeds near their neighbors' genetically engineered crops.

Nell Newman, Paul Newman's daughter and founder of Newman's Own Organics, a division of Newman's Own, complained that "extensive planting of GMO crops in the U.S. has become the bane of my existence and threatens my livelihood in the organic food industry." Because of the potential contamination of organic crops, the company is now having to test all of its ingredients—and absorb the enormous costs incurred.

"For a company that donated all of its profits to charity," Newman said, "it means that I'm now wasting my charity dollars trying to figure out whether or not Monsanto has cross-contaminated any of my raw ingredients. This year one of my corn growers spent over sixteen thousand dollars testing for GMOs to insure that her crop had not been cross-pollinated. My question to the FDA is: in the coming years, what safe haven will I have to plant my organic crops in, and who will cover the liability and losses that are bound to occur?"

If an organic wholesaler or retailer detects even trace amounts of genetically modified corn in a food product labeled as containing organic corn, it can render an entire lot of food worthless for the grower, the processor, and the distributor. The first documented case of contamination involved Terra Prima, a small organic company in Hudson, Wisconsin. Terra Prima supplies Apache Brands organic chips to stores in seven countries in Europe. In December 1998, after a European retailer performed tests on its organic tortilla chips and found they contained gene-altered corn, Terra Prima was forced to recall more than 87,000 bags of the chips.

Melodi Nelson, vice president of Terra Prima, traveled to Oakland to appeal to the FDA to rescind registrations of genetically modified organisms used in agriculture. When pollen from such crops contaminates organic crops, she charged, it threatens the rights of people who consciously choose to raise and eat natural foods. "We have the right to not have our farms contaminated and our livelihoods taken away from us. We have the right to know what we choose to eat and feed our children," she told the FDA. "We have the right and obligation to be skeptical. What is at stake is the safety and the security of our food source around the world."

Robert Cannard of Cannard Farms warned the FDA that uncontrolled pollen drift from gene-altered crops was "the death knell of any organic industry in this nation. It doesn't matter what levels you talk about as being permissible, they'll soon be exceeded."

Cannard, a Sonoma County organic farmer, had filed a proposed ballot initiative with the state attorney general in October 1999 to require labeling of genetically engineered food. But supporters of the California Right to Know Genetically Engineered Food Labeling ballot initiative fell short of the 419,260 signatures needed to place the measure on the November 2000 ballot.

Roy Fuchs, director of regulatory science for Monsanto, spoke at the hearing, vouching for the safety of his company's products. Fuchs asserted that more than eighteen hundred analyses were conducted with Roundup Ready soybeans to establish their safety. "The protein used to confer tolerance to Roundup herbicide is a member of a protein family which occurs in every plant, every bacterium, and every yeast, all of which have been consumed safely for centuries. In addition, detailed safety studies were conducted with each of these proteins to confirm their safety," he said. "As a developer of these products, we're committed to ensuring their safety."

Kara Cosby, media relations manager with the International Food Information Council, told the FDA that the government's policies on

the new foods should not be driven by opinions that were less than informed. "Our consumer surveys have consistently shown that consumers with the highest level of education have been most likely to
support food biotechnology. When consumers have solid, science-
based information, they make the right decisions," Cosby said. "A
majority of consumers support the benefits of food biotechnology
when they are clearly explained."

Paul Bettencourt, a cotton grower from Fresno County in the San
Joaquin Valley, praised the performance of his Bt cotton crop. He said
the Bt cotton that he planted for the first time in 1999 was better for
the environment than conventional chemical pest control. "Here I had
a product that saved me at least two applications of pesticide for the
same amount of bug control, and we're getting criticized for it," he
told the FDA.

Mike Phillips, an official with the Biotechnology Industry Organization, insisted that BIO supports the right of consumers to "make
informed choices regarding the foods they eat." Like Agriculture Secretary Dan Glickman and other proponents of biotech foods, he argued that organic foods should provide the niche for U.S. citizens who
wish to avoid eating genetically engineered foods. "Choice is provided. If a consumer wants to buy food not derived from enhanced
crops, they can buy organic and also pay the higher cost." Phillips,
who previously served on the food and agriculture board of the National Academy of Sciences, offered support for the FDA's 1992 labeling policy. "Changing policy to require special labeling for foods
derived from biotechnology could impact significantly consumers'
perception of the safety of these foods and undermine the confidence
consumers presently have."

Janet Brown, a California certified organic farmer and founder of
the Marin Food Policy Council, described a resolution passed by the
Berkeley City Council a week earlier. It supported the establishment of
a federal ban on "growing, disseminating and marketing products genetically engineered until they have been proven safe for human consumption." Brown told the FDA that the Marin council was worried
about "the permissive atmosphere which allows an essentially unregulated and disturbingly disintegrated science of genetic engineering to
operate in the absence of liability constraints."

The Berkeley resolution, adopted December 7, 1999, was one of a growing number of local measures cropping up around the country to rid localities of genetically engineered foods and plants. In March 2000 the Boston City Council passed a resolution urging the federal government to require labeling of genetically manipulated foods. It called for "a moratorium on the production of any more of these foods until acceptable testing systems are in place." The resolution went on to declare March 26, 2000, "You are what you eat day in the City of Boston."

On May 18, 2000, the city council of Austin, Texas, passed a resolution calling for mandatory labeling and a moratorium on new genetically manipulated foods. In July the San Francisco City Council passed a more elaborate resolution, urging the city and county to give preference to certified organic food vendors and urging the federal government to label and test genetically engineered foods and, in addition, to "assign liability to the commercial developers of G.E. technology."

Cleveland mayor Michael White signed an emergency resolution on August 17, 2000, urging the federal government to require labeling and place a moratorium on the production of new genetically manipulated foods until safety systems were in place. One week later the Minneapolis City Council passed a resolution patterned on the San Francisco measure. Also in August 2000 the Boulder Open Space Department approved a policy barring genetically modified agricultural crops from city lands, citing the lack of information on their long-term ecological impact.

Organic producers are not the only victims of biological pollution from genetically engineered pollen. Conventional farmers who attempt to grow traditional, nonengineered varieties have found it increasingly difficult to guarantee the purity of their crops. George Naylor, a corn and soybean farmer in Green County, Iowa, harvested more than ten thousand bushels of corn and seven thousand bushels of soybeans in the fall of 1999. He hopes the grains are not contaminated with genetically engineered organisms—but he can't be sure.

"I've never raised GMO crops," Naylor said at a press briefing in

Washington, D.C., sponsored by the Consumer's Choice Council in February 2000. The Council is made up of more than fifty diverse consumer and environmental groups advocating labels on gene-altered foods, including the Sierra Club, National Audubon Society, Co-op America, and Physicians for Social Responsibility. "Unless there was some compelling reason to raise them, I would be opposed to taking a gene from another organism and putting it into our crops, considering the three billion years of evolution that's involved. I've tried to protect what I have grown so that it would be free from GMO grains."

George Naylor could not control his neighbor's crop choice, however. Because his neighbor raised Bt corn, Naylor assumes that some pollen drifted from those crops onto the edges of his own cornfields. To get rid of the corn that might have been contaminated, Naylor combined a hundred rows of corn off the sides of each of his fields. He paid for extra drying and hauled the harvested corn to the local grain elevator, where it was blended into the general grain supply.

He stored the rest of the corn in bins on his farm. "I'm hoping that maybe it's non-GMO," Naylor said. But news reports in seed industry magazines have troubled him. "The seed industry now has a standard of a thousand feet, which would be four hundred rows. I've combined off a hundred rows, thinking that should be good enough, but now the industry's saying four hundred rows, so I can't be sure, really, what I've got in my bins."

Naylor's farm is a mile long and a half-mile wide. Using these industry guidelines for isolation, he calculated he would end up having only fifty acres in the middle of his own farm on which he could raise corn. "There are no rules," Naylor told reporters. "There's nothing to say that the farmer is obligated to distance his GMO crops from my crops. He normally plants it where he wants, which is six feet from my crop.

"Also, I don't know how pure the seed was that I bought in the first place," Naylor continued. He decided to take extra care with his 2000 crop. He contacted his local seed dealer for Pioneer International and asked, "Can you certify that the seed I'm buying this year is non-GMO when I pick a non-GMO variety?" The seed dealer called his supervisor to get an answer. "The word came back that no, they would not certify that. So in effect, I don't know what I'm planting," said Naylor, his voice rising with emotion.

"This is really a nightmare out there. We don't know what we're planting. We don't know what we're putting into the food supply."

George Naylor and his wife, Peggy, joined three other U.S. growers and one French farmer in launching a class-action lawsuit against Monsanto. The lawsuit, filed December 14, 1999, charges Monsanto with rushing genetically modified corn and soybeans to the marketplace without adequately testing them for risks to human health and the environment. It alleges that Monsanto and a handful of other life sciences companies, termed "co-conspirators," have created a cartel aimed at controlling the U.S. seed market. Named as co-conspirators are Delta and Pine Land Company, DuPont, Pioneer Hi-Bred International, Dow Chemical Company, Mycogen Corporation, Novartis International, AstraZeneca, Garst Seed Company, and AgriPro Seeds.

Monsanto strenuously refuted the accusation. "Farmers today have more choices of high quality seed than ever before," the company said. "It was the farmer demand for better seed that brought biotechnology to growers worldwide in 1996." Voicing unwavering belief in biotechnology, the company added: "We are committed to the farmer's right to have access to high-quality seed choices that allow them to continue in the outstanding progress they have made in producing more food with fewer inputs on less land."

Meanwhile, the farmers believed that Monsanto and other biotech seed companies should be held accountable for any environmental damage caused by their crops, for financial damage to farmers resulting from cross-pollination, and for the loss of export markets for American growers who have been unable to sell their crops abroad because of concerns about health and safety.

"It is a concern of every farmer that we've spoken to that Monsanto has put these products out there and left farmers out to dry," Elizabeth Cronise, an attorney with the Washington, D.C., law firm of Cohen Milstein Hausfeld & Toll, one of three law firms involved in the case, said at the press briefing.

The lawsuit accuses Monsanto and its co-conspirators of requiring farmers to sign technology user agreements when they purchase genetically modified seeds and then using those agreements to intimidate growers into complying with the cartel's scheme. Before the advent of the new crops, farmers had never had to pay these "technology fees." Farmers could save, trade, and plant seeds from the crops they harvested. Monsanto and other biotech seed companies, the brief charges, have filed more than 475 lawsuits against farmers for supposedly violating technology user agreements by saving genetically modified seeds. In addition, the companies have attempted to enforce the terms of technology user agreements against farmers who never entered into any such agreements at all. Monsanto set up a "snitchline" on which neighbors could call to report local farmers who they suspected might be growing its patented seeds without paying the technology fee.

The most celebrated case testing the legality of Monsanto's enforcement practices is that of Percy Schmeiser, a canola farmer in the small rural town of Bruno, Saskatchewan. Schmeiser is a seventy-year-old farmer who has been growing canola in Canada since the early 1950s. He saves and reuses his own seed from year to year. Acting on a tip received on the company's "snitchline," Monsanto sent agents to Schmeiser's farm and removed some of his canola plants for DNA testing. In 1998 Monsanto charged the Saskatchewan farmer with deliberately growing Roundup-resistant canola from seeds saved in 1997, thereby infringing on the company's patent.

Monsanto sought damages that included $13,500 for technology fees, at $15 an acre, plus $105,000 in profits that the company believes Schmeiser made on his 1998 crop, and $250,000 in legal fees. Schmeiser contends that if he had negotiated an out-of-court settlement early on with Monsanto, as most farmers have, the company might have settled just for the technology fee. But he decided to dig in and fight for the right of farmers to save their seeds. His case has grabbed headlines around the world.

Schmeiser acknowledges that Roundup-resistant canola plants, containing Monsanto's gene, were growing on his property, as Monsanto alleges. But he steadfastly insists the plants with Roundup resistance made their way onto his farm against his wishes. "This is the result of contamination of Mr. Schmeiser's seed supply, and not by design. Monsanto cannot control, and has never tried to control, the spread of its gene around the countryside," his lawyer argued in a trial brief. Schmeiser said that he had developed a superior variety of canola in his many years of cultivating the crop, and that because of the contamination by Roundup Ready canola on his farm, he had been required to buy all new seed in the 1999 growing season.

Schmeiser's attorney argued in a court trial in June 2000 that

Roundup-resistant canola routinely finds its way onto fields where farmers never intended to grow it. The plants fall off trucks passing by farm fields, and their pollen is carried on the wind or by insects onto neighboring fields. Roundup Ready canola was rapidly adopted by Canadian farmers, and by 1999 it accounted for a major share of seeded acres of canola in western Canada. The gene for herbicide resistance, which Monsanto patented, is dominant. This means that when Roundup Ready pollen fertilizes conventional canola plants, the resulting plants are resistant to Roundup.

Contamination of farmers' fields and roadsides with Monsanto's canola plants is so widespread, Schmeiser claimed, that soon it will be hard to find a canola field in western Canada without the company's gene. He argued that if Monsanto had maintained control over its invention, the company might have maintained its exclusive rights under its patent. But given that Monsanto never controlled, or attempted to control, the spread of its invention into the environment, the company lost its right to assert patent infringement claims against people on whose property plants resistant to Roundup sprang up. Monsanto, the brief argues, "cannot on the one hand unleash self-propagating matter uncontrolled into the environment and then claim exclusivity wherever it invades."

At issue, ultimately, is the question of whether a farmer can save and plant the seeds from crops grown on his own property. "In my case, I never had anything to do with Monsanto, outside of buying chemicals. I never signed a contract," Schmeiser pleaded. "If I would go to St. Louis and contaminate their plots—destroy what they have worked on for forty years—I think I would be put in jail and the key thrown away."

To millions of farmers around the world, Percy Schmeiser has become a folk hero. In October 2000 he traveled to Delhi, where he was awarded India's Mahatma Gandhi Award. The award is given in recognition of work done for the betterment of mankind in a nonviolent way.

While Schmeiser might have prevailed in the arena of public opinion, he soundly lost in the court of law. On March 29, 2001, Justice Andrew McKay upheld the validity of Monsanto's patented gene for Roundup resistance in canola. The ruling, by the Federal Court of Canada, dismissed Schmeiser's argument that Monsanto had lost control over the gene by allowing it to spread throughout farmers' fields in western Canada. Schmeiser appealed in June 2001, and the case is still pending at this writing.

With genetically modified seed and pollen increasingly contaminating the fields of organic and conventional growers, the National Farmers Union (NFU) in Ottawa called on the Canadian government to hold agricultural biotechnology corporations financially responsible for genetic pollution. "It may soon become impossible to certify canola as organic because no one will be able to guarantee that it does not contain genetically engineered seeds," the union warned in a May 1999 statement.

"If this continues," said Stewart Wells, Saskatchewan coordinator of the NFU and himself an organic farmer, "once wheat, barley, lentils, and other crops are genetically engineered, I won't have anything left to grow. For organic farmers and the hundreds of thousands of consumers who choose organic food, this is an extremely serious issue."

Conventional Canadian farmers have already lost markets in the European Union because of widespread genetic pollution of their canola crops. "The loss of the European market will have a significant cost to all farmers," said Wells. "Monsanto and other companies who engineer these crops reap the profits of their work. It is also fair that they pay the costs which their products create for other farmers."

Gary Goldberg, CEO of the American Corn Growers Association, took up the cause of conventional farmers in the U.S. corn belt. "Farmers have been caught in the middle of this debate between seed dealers, chemical companies, agro-processors, grain traders, foreign consumers, and U.S. trade policy," he said at a February 2000 meeting in Washington, D.C., sponsored by the Worldwatch Institute, a small research organization in Washington, D.C. Like many organizations, it developed an interest in biotech foods only after the issue heated up in 1999, following publication of John Losey's monarch butterfly study. "Through advertisements and farm publications, radio and television, the farmer has been presented with planting options that never address consumer resistance or questionable testing results. Certain other farm organizations have received so much corporate financial support from biotech interests that they have been compromised. And this administration has pushed an arrogant trade agenda that attempts to dictate what our foreign customers should buy—and threaten them with trade sanctions if they refuse."

Goldberg insists he does not oppose GMO crops but is simply responding to the demands of consumers in the United States and abroad. "From beer brewers and millers in Japan, to the major grocery store chains in England, to the largest tortilla maker in Mexico, GMOs are being rejected," he said. "So now we will be left with bins full of GMOs and limited opportunities of where we can market them."

American farmers who plant nonengineered crops should have their seeds tested to verify that they are free of genes from bioengineered varieties, Goldberg said. Evidence of seed contamination had already turned up in Switzerland and Germany. Certain varieties of seed corn that had been sold as nonengineered were contaminated with between 0.1 and 0.5 percent unapproved genetically modified corn, Swiss and German authorities discovered in May 1999. Fields where the illegal corn was detected were destroyed and the farmers were compensated.

Seed companies should be required to perform tests to verify that nonengineered seeds are pure, Goldberg said, but they cannot be relied upon to do so. He urged farmers not to wait for the companies to test. "Farmers who are seeking to sell to the non-GMO markets must protect themselves by getting their seeds tested. If we are to maintain our export markets and possibly gain non-GMO premiums, we have to take steps to guarantee the purity of our crops."

Six months after running three very public meetings on genetically engineered foods, the FDA contracted with a consultant to do some quiet research into the American public's perceptions. In May 2000 the agency hired ORC/Macro International to run a series of focus groups to discover how American consumers feel about genetically engineered foods. During the first meeting, which was held in Calverton, Maryland, a moderator gave paid volunteers some basic information about gene-altered foods. Many of the participants were college educated; all had at least a high school education. Few had any idea that more than 20 percent of all U.S. corn and half of its soybeans were genetically engineered. They were outraged when they learned that more than 60 percent of the foods on the grocery store shelves contained genetically engineered ingredients.

"After being presented with a factual account of the extent to which certain grain crops in the U.S. are being produced from bio-engineered seed and the extent to which bioengineered ingredients are present in processed foods, most participants expressed great sur-prise," an official FDA report dated October 20, 2000, stated. "Even among participants who considered themselves well-informed about biotechnology, many registered amazement. The typical reaction of participants was not one of great concern about the immediate health and safety effects of unknowingly eating bioengineered foods, but rather outrage that such a change in the food supply could happen without them knowing about it."

Following the initial focus group, the FDA held eleven more, host-ing them in Maryland, Burlington, Vermont, Seattle, and Kansas City, Missouri. Across the country, consumers had the same angry reaction to learning how pervasive biotech food has become in the U.S. food supply.

"Some participants remarked that bioengineered foods have been 'snuck in' to the food supply. They were mainly disturbed by the lack of public information and public input to a major development in the quality of their food supply," FDA officials reported. "Some partici-pants saw this as evidence of a conspiracy to keep consumers in the dark; that is, the rationale for not informing the public must be that there is something to hide."

From Maryland to Missouri, citizens shared their worries about "unknown long-term health consequences that might be associated with the technology but which cannot be anticipated based on current science or knowledge." This concern was nearly identical to that voiced by European citizens in 1997 and 1998—concern that prompted Euro-pean officials to adopt the precautionary principle in regulating biotech foods.

Focus group members drew parallels between the introduction of these foods and other technological innovations in modern agriculture, including the use of pesticides, growth hormones and antibiotics to promote animal growth, and tomatoes bred for transportability rather than taste. "In each case, participants saw a technological innovation that was introduced mainly for the sake of producers/distributors, with little apparent benefit to the consumer. Such innovations are seen as being approved by scientists and regulators, but later found to have unanticipated long-term health effects," the FDA officials reported.

The FDA's purpose in conducting these meetings was ostensibly to help the agency develop a new policy for genetically engineered foods. The 1992 policy was now nearly universally viewed as woefully deficient. Even food industry trade groups such as the National Food Processors Association and the Grocery Manufacturers of America had joined with consumer groups in telling the agency that company consultations should be mandatory—not voluntary—prior to the introduction of gene-altered foods into the food supply. But the trade groups and biotech companies refused to budge from their vehement opposition to mandatory labels. Nearly all public opinion polls, with the exception of those conducted by the food industry, have found that members of the public overwhelmingly support labeling. Hence it should have come as no surprise to the FDA that the men and women in the focus groups also called for labeling. "Virtually all participants said that bioengineered foods should be labeled as such so that they could tell whether a given food was a product of the new technology," the FDA report stated.

Participants complained that there was no way to tell which foods had been genetically engineered and "were concerned about the potential for unknown long-term effects of the technology, in particular health effects." Some focus group participants said they would avoid genetically engineered foods for health reasons; others, "to send a message" to the companies making the products.

As the focus groups wore on, and consumers were told about kinds of genetically engineered foods that might be expected to come along in the future—including vitamin A–enriched rice and growth-hormone salmon—they "began to see the value of having more information [on a label] than mere disclosure of whether products are produced by food biotechnology," the FDA officials reported.

In the course of holding three public meetings and twelve focus groups, the FDA had spent considerable taxpayer money canvassing Americans' views on genetically engineered foods. The message was unequivocal: when Americans learned about genetically engineered foods, they wanted labels. But the food industry and the biotechnology lobby intensified their efforts to fight mandatory labeling, hoping, and perhaps even believing, that the issue would fade into obscurity before most consumers even learned they were eating something new.

TWELVE

GOLDEN RICE

A fter the collapse of the WTO talks in Seattle and the rebuff of American interests in Montreal, the biotechnology industry needed a miracle to resuscitate the image of genetically engineered food. That miracle came in the form of a tiny golden grain of rice. Following nearly a decade of laboratory work, Ingo Potrykus, a professor at the Swiss Federal Institute of Technology in Zurich, succeeded in accomplishing a daunting feat: he engineered the pathway for vitamin A into rice. His genetically engineered rice held out the potential for preventing disease and death among millions of the world's poorest people—as well as rekindling hope in an industry in desperate need of good news.

Deficiency of vitamin A is a scourge among people in the developing world. Each year about half a million children go blind and more than a million suffer a host of diseases for lack of sufficient vitamin A in their diets. Meat, fish, and animal products are good sources of vitamin A, as are carrots and leafy green vegetables. But for too many of the world's poorest people, meat and vegetables are luxury items that seldom grace their palates. Half of the global population depends on rice as a food staple—and many of them have little to add to their bowls of rice.

Normal rice, *Oryza sativa,* produces beta-carotene, which our bodies convert into vitamin A, in its leaves but not in the pearly white grains that people eat. Potrykus and his colleague Peter Beyer of the University of Freiburg in Germany transferred genes from the daffodil and a strain of bacteria into rice plants. The result was "golden rice"—a yellow-colored rice that contains beta-carotene.

At a science meeting sponsored by the journal *Nature Biotechnology* in London in November 1999, Potrykus reported that a typical Asian diet of three hundred grams of the yellow rice alone would provide a person with the necessary daily dose of vitamin A to prevent deficiency. For Potrykus and his backers at the Rockefeller Foundation, the possibility that gene-altered rice might one day relieve many children of disease and blindness was nothing short of miraculous. In late 1999 the biotech rice plant was still just a laboratory phenomenon— no one could be certain that it would thrive in the field as it had in the hothouse. Also, scientists had to move the genes into a different variety of rice. Potrykus and his colleagues had inserted the novel genetic material into japonica rice, which is short-grained, because it proved easier to alter than other varieties. Breeders were now in the midst of trying to cross the desired trait to long-grained indica rice varieties, which are more popular in Asia. Once breeders succeed in introducing the trait into long-grained rice, scientists will then have to perform environmental and toxicity tests on the golden rice before it can be grown and eaten by people. Potrykus estimated that within five years, vitamin A rice could be planted and harvested by poor farmers.

Assuming the technical, human safety, and environmental issues associated with the new transgenic crop can be resolved, golden rice will need to overcome cultural barriers as well. Peasants in Asia might not accept the rice's yellow color, particularly if golden rice becomes associated with poor people while wealthier people continue to eat white rice. Nonetheless, the scientific breakthrough was significant, and it portended a bright future for the biotech crop.

Ingo Potrykus and his colleagues also were able to engineer pathways for iron into rice plants. An estimated 1.4 billion women around the world suffer from iron deficiencies. Rice contains less iron than other cereal grains, and it is also high in phytate, a compound that inhibits the absorption of iron in the intestine. Potrykus tackled the problem by inserting a fungus gene into the rice plant that produces an enzyme called phytase. The enzyme eliminates the unwanted phytate from the rice. He also added a ferritin gene to the rice, to increase the plant's iron content.

Apart from its potential to alleviate disease for some of the world's neediest people, golden rice carried important symbolic value. Here at last was a gene-altered crop that gave biotechnology a humanitarian face. Unlike pesticide-producing corn and herbicide-tolerant soybeans, which are fraught with ecological and long-term safety uncertainties, golden rice seemed to hold out the possibility of delivering tangible health benefits to malnourished people in developing nations. Instead of enriching multinational corporations, as do crops bioengineered to be used hand in glove with brand-name chemicals, vitamin A rice is intended to be given away freely to rural peasant farmers.

Golden rice soon became the focal point of an impassioned debate over the future of agriculture and the security of the world's food supply. In the months following the announcement of Potrykus's new strain of rice, it was hard to attend a food policy meeting where U.S. politicians and biotechnology industry executives did not extol the awesome potential of vitamin A rice to alleviate malnutrition and suffering.

In January 2000 Missouri senator Kit Bond and Roger Beachy of the Danforth Plant Science Center traveled to Thailand as U.S. biotech emissaries. Consumers in Thailand, following the path of their neighbors in Japan and Korea, were growing increasingly wary of genetically modified foods. Senator Bond used the discovery of vitamin A rice to illustrate the benefits that biotechnology might deliver to Thai farmers. "One of the things that we have discussed with the university scientists here is the addition of a gene for beta-carotene, which would add vitamin A to the wonderful rice that is produced in Thailand. This vitamin A could be a very valuable tool in assuring a well-rounded diet for children in less developed countries," Bond told Thai reporters during a roundtable discussion at the U.S. Embassy in Bangkok on January 21, 2000.

The following month, a State Department official seized on the potential promise of vitamin A rice to justify the need for increased public funding of biotechnology research around the globe. "I would point, as an example of the type of promise that this technology offers, to the recent work on vitamin A enrichment of rice, so-called golden rice," David Sandalow, assistant secretary of state for oceans and international environmental and scientific affairs, told reporters and policymakers at the Worldwatch Institute briefing on genetically engineered food in Washington, D.C., on February 12, 2000. "There are opportunities to address critical nutritional deficiencies in the developing world with this type of technology—not just vitamin A, but other micronutrient deficiencies. In order for this technology to be useful, we need money to do research, particularly on tropical crops."

At the same Worldwatch Institute briefing, Per Pinstrup-Andersen, director-general of the International Food Policy Research Institute, said the public sector must invest in any avenue of research—including biotechnology—that might lead to solutions for the world's millions of hungry. "The context within which poor people in developing countries operate—and seventy percent of them are in rural areas—is frequent droughts, frequent attacks by insects and plant diseases as well as animal diseases, very low soil fertility, and soil mining," he said, adding that the possibility of increasing micronutrient content of staple foods—including vitamin A, iron, and zinc—offered a "glimpse of light" on the horizon for rural farmers.

While acknowledging that biotechnology is not a "silver bullet" to cure all the ills that plague rural farmers and poor city dwellers in developing nations, Pinstrup-Andersen, who hails from Denmark, argued that it could provide part of the solution. "We do not think we should withhold any potential solutions to the problems facing poor people in developing countries, because this is not a philosophical debate for them. This is a matter of life and death."

He also argued that wealthy Western consumers should not be allowed to decide whether farmers in developing countries should have access to the techniques for genetically altering food crops.

Susan McCouch, professor of plant breeding at Cornell University, was among the many scientists at the annual meeting of the American Association for the Advancement of Science in Washington, D.C., in February 2000 who pointed to vitamin A rice as a biotechnology product that offers immediate assistance to people in need. "People who suffer from malnutrition generally lack essential levels of micronutrients because they lack the purchasing power to obtain sufficient diversity in their diet," she said.

McCouch likened the use of genetically engineered yellow rice in developing nations to the U.S. practice of fortifying milk with vitamin

D, salt with iodine, or orange juice with calcium. In addition to its potential nutritional utility, she suggested, golden rice might help to persuade leery Western consumers that agricultural biotechnology is valuable. Opinion surveys showed that people are initially uncomfortable with the idea of using biotechnology to transfer genes between organisms, she noted, but such reservations could be overcome if people perceived a particular ethically or morally persuasive benefit.

The director-general of the UN Food and Agriculture Organization came out in support of biotech crops, stating that genetically modified food will be needed in the long term to fight hunger. "We cannot deprive ourselves of the potential to have crops that require less pesticide, need less nitrogen and phosphorus to grow and offer poor people improved nutrition," Jacques Diouf told the *Financial Times* on June 28, 2000.

The following month President Clinton touched on the significance of vitamin A rice at the G-8 economic summit in Okinawa, Japan. Addressing the question of whether genetically modified food is safe, Clinton said, "All the evidence that I've seen convinces me, based on what all the scientists now know, that it is. But, of course, every country has to deal with that. But just for example, if we could get more of this golden rice, which is a genetically modified strain of rice especially rich in vitamin A, out to the developing world, it could save forty thousand lives a day, people that are malnourished and dying."

The week after Clinton's endorsement of vitamin A rice, *Time* magazine featured Ingo Potrykus on its cover. "This rice could save a million kids a year," bold letters on the cover proclaimed, "but protesters believe such genetically modified foods are bad for us and our planet."

Where advocates of genetic engineering saw a golden miracle, however, critics spotted fool's gold. Some environmentalists viewed vitamin A rice as little more than a public relations ploy being exploited by the biotech industry to revive public support for genetically engineered foods in Europe and much of Asia. Greenpeace challenged the notion that golden rice would prevent vitamin A deficiencies among the world's poor. The environmental group based its critique on a study by Xudong Ye, published in the January 14, 2000, issue of *Sci-*

ence. According to the study, the "best line" of genetically engineered golden rice produces lower levels of beta-carotene than scientists previously suggested. A woman would have to eat about nine kilograms of cooked rice to get the required dietary intake of vitamin A—twelve times the normal intake of rice. Based on the findings of the scientists, Greenpeace calculated that three ordinary servings of golden rice would provide only about 10 percent of the vitamin's required intake.

Greenpeace urged international aid agencies to continue funding three effective strategies to fight vitamin A deficiency: handing out oral doses of vitamin A to mothers and children on an emergency basis, fortifying foods such as butter with the vitamin, and promoting access to cheap, varied dietary sources of vitamin A.

Potrykus responded that nutritionists have suggested his golden rice could potentially meet between 20 and 40 percent of the daily allowance of vitamin A. He and other scientists acknowledged, however, that golden rice alone would not solve the problem of malnutrition.

Other critics believe that even if crops like golden rice can be made to work, they will encourage monoculture in the developing world and lead to another set of health and environmental problems in the future. They argue that agriculture policies should encourage farmers around the world to adopt sustainable farming practices that build up soils rich in microorganisms and encourage the cultivation of dozens of varieties of food plants, rather than foster reliance on one or two vitamin-enriched crops.

Martha Crouch, who worked for years as a plant geneticist at Indiana University in Bloomington, charged that the developers and promoters of golden rice are thinking about food as "a bag of chemicals." People taking this perspective have little regard for how people get the nutrients and energy they need, she said, "whether from rice or vegetables or a mixture of foods." Taken to its extreme, this system would reduce food to "human Purina chow," she told participants at a "Biodevastation4" meeting sponsored by antibiotechnology activists in Boston on March 25, 2000. People would be forced to eat "human kibble" that contains all the necessary ingredients.

One of the many problems with this approach to solving hunger problems, Crouch argued, is that people's diets are linked to the environment through agriculture. "If you have all of your staple foods with everything that you need in them, then you only have to grow corn, or you only have to grow rice, which means that there's a monoculture on the land, which means that you've created a uniform environment where only a few species will find it attractive to live," she continued. "If you were growing a hundred or a thousand different species of vegetables and fruit trees, bushes, grasses and legumes," hundreds or even thousands of species would find it attractive to live in and around farms.

"So you're making a monodiet into a monoculture, which decreases biodiversity, because agriculture is part of nature, and our diet is linked to agriculture," she concluded. "Food and agriculture go together."

Vandana Shiva, an Indian scientist and fiery orator, is among the world's sharpest critics of plant genetic engineering. Like Crouch, she is a strong opponent of monoculture and an advocate of organic agriculture.

The antidote to "large rows of monoculture," Shiva believes, is having polyculture and cover crops. "You don't live by corn alone. You need the squash and you need fruits." Successful polyculture can be found among farmers in Java whose gardens can contain as many as 600 species, or in Thailand, where 230 species are found in one garden. "In Nigeria the home gardens cultivated by women on two percent of the land provide fifty percent of the nutrition," she told an audience gathered at the U.S. Capitol on June 29, 2000, for a debate on the role of biotechnology in fighting world hunger.

Shiva noted that Roundup Ready crops not only promote monoculture but actually show a yield drag. She also highlighted the success of certain organic farmers in Guatemala and Honduras who have increased their output significantly by switching to organic techniques.

Channapatna Prakash took strong exception to Vandana Shiva's vision of a productive organic-farming and sustainable-agriculture system based on indigenous knowledge in the developing world. "I'm frankly sick and tired of hearing those kinds of arguments, because I grew up seeing what local knowledge is," Prakash responded during the debate at the capitol. "Local knowledge is losing one-third of your children before they hit the age of three. Is that the kind of local knowledge you want to keep reinforcing and perpetuating? What is sustained is misery and disease."

Prakash ridiculed the notion that organic techniques could sustain the global population. If the whole world went organic, he said, we would need six billion cows to produce enough manure. "Eighty percent of farmers in India are organic. If you give them a dollar, the first thing they'll do is go buy some pesticide so they will make a better yield."

Genetic engineers can use their skills to develop new varieties of plants, Prakash argued. "Biotechnology is just one other tool that we have that is more precise, it is more flexible and more expedient." As to questions about the safety of gene-altered food, Prakash said that "270 million Americans have been eating this soybeans and corn," and regulatory authorities have found there's nothing wrong with them.

During the same debate, Dennis Kucinich, who introduced the first mandatory labeling bill in Congress, argued that genetically engineered crops will not help solve the paradox of world hunger. "The world is a cornucopia of food, yet people are still hungry in all nations, including this one," he said. "Even if genetically engineered food has some yet-to-be-discovered intrinsic benefit, this benefit certainly does not override the people's right to know—and the necessary assurance that the food is safe."

Some proponents of biotechnology have transparently used Ingo Potrykus's creation as a tool to blunt Western critics of Bt corn and Roundup Ready soybeans. They often imply that it is immoral for wealthy consumers in Europe or Japan or North America to stand in the way of the development of a technology that might one day help millions of diseased and malnourished children in developing nations. The assumption is that developing nations could not support the infrastructure and expertise needed to design and grow genetically engineered crops suited to their particular needs if people in the West refused to grow, import, or consume them. Such arguments can easily prick the conscience of well-fed Western consumers who may not want to eat genetically engineered foods but certainly do not want to be responsible for increasing the misery of the world's less fortunate people.

This argument rankles Tewolde Berhan Gebre Egziabher, who is an official with the International Institute for Sustainable Development in Addis Ababa, Ethiopia, and served as spokesperson for Africa and the "Like-Minded Group" in the negotiations for the Biosafety Protocol. He wrote an angry letter to British Channel 4 Television after watching a documentary that had suggested the average English housewife's resistance to genetically engineered food would prevent the development of biotechnology, the very technology needed to eliminate malnutrition in Africa. Egzhiaber said he was "appalled at the use made of the poverty of the rural people of the South to justify genetically modified food to Northern consumers."

It is not the shortage of food that is a problem in much of the developing world, he noted, but rather the lack of adequate food distribution. The best way to combat hunger in such conditions, he argued, is by improving access to food and increasing local food security rather than by developing genetically engineering crops.

However, Florence Wambugu, who is a passionate and articulate advocate of the benefits that agricultural biotechnology might hold for African farmers, sees a silver lining in the consumer backlash to biotech foods in Europe. She believes that the confrontation between environmentalists and biotechnology advocates has changed public and private sector research for the better. "Private-sector multinationals previously aloof to public opinion and the needs of the poor have discovered a new compassion. Many now work directly with the public sector in developing countries or else have launched their own charitable foundations to do so," she wrote in *Modifying Africa: How Biotechnology Can Benefit the Poor and Hungry, a Case Study from Kenya.*

In the early 1990s Wambugu received a scholarship from the U.S. Agency for International Development and took up a fellowship in biotechnology at Monsanto's Life Sciences Research Center in St. Louis. She and colleagues developed the first genetically modified sweet potato designed to resist feathery mottle virus, a scourge for African farmers. In 1994 Wambugu returned to Kenya to become director of the AfriCenter of the International Service for the Acquisition of Agribiotechnology Applications.

In addition to learning genetic engineering techniques from Monsanto, Wambugu clearly absorbed valuable lessons from the company's failed marketing approach in Europe. African scientists are taking a cautious approach to the introduction of genetically modified crops, she said, in hopes of forging public support for biotechnology and avoiding mistakes made by developed countries. "The fact that neither the public nor the private sector is pushing biotechnology products means that the public feels it has a choice in the matter and is not having GM organisms 'stuffed down its throat,' as the complaint has been in Europe," she noted.

The United Nations Development Program (UNDP) has warned, nonetheless, that vocal critics in Europe and the United States are slowing down research and could prevent biotechnology advances from reaching the world's poorest farmers. While biotechnology offers the potential to develop crops for marginal ecological zones, according to Sakiko Fakuda-Parr, author of a UNDP report issued in 2001, 98 percent of the word's transgenic crops were under cultivation in just three countries, the United States, Canada, and Argentina.

While scientists and food policy experts debated the potential future benefits that biotechnology might bestow on the developing world, farmers and ecologists in North America were discovering some oddities in the gene-altered crops growing across wide swaths of the continent.

In February 2000 the *Western Producer* reported that a Canadian farmer discovered a strange canola plant in his northern Alberta field. The plant was resistant to three different herbicides—Monsanto's Roundup, Aventis's Liberty, and Cyanamid's Pursuit. In 1997 the farmer had seeded canola resistant to Roundup on one side of a county road and canola resistant to Liberty on the other. He planted the remainder of his acreage with a seed tolerant to Pursuit. The following year, "volunteer" weeds resistant to Roundup sprang up where the farmer had not planted the variety. The next year volunteers sprang up that had genes for resistance to all three weed-killers. Canola experts said it was the first known example of "stacked" biotech traits occurring in nature.

The bright yellow flowers of the canola plant are ideal for spreading pollen over great distances, on the wind and on insects. Volunteer canola—plants that crop up in fields the following year—are common weeds in western Canada. Farmers who wish to plant a rotation of barley or wheat routinely apply an herbicide to the field to kill the unwanted canola "weeds." But the only reliable way to kill the new triple-resistant canola weeds was to apply "2, 4-D," a strong chemical herbicide that farmers used much more widely before the advent of "safer" herbicides such as Roundup.

On May 28, 2000, British newspapers reported that genes from genetically engineered plants had transferred into bees. A German scientist found that a gene from genetically engineered canola had made its way into bacteria and fungi in the gut of honeybees. Pollen, which is high in protein, provides the natural diet of baby bees. The scientist, Hans-Hinrich Kaatz of the Institute for Bee Research at the University of Jena, performed his experiment using a variety of canola that had been engineered by Aventis to resist the company's herbicide glufosinate.

Also in May, Monsanto revealed astonishing new information about Roundup Ready soybeans. Four years after they were first harvested for human consumption, Monsanto told the USDA that company scientists had just discovered two "unexpected" DNA fragments in Roundup Ready soybeans. The previously unobserved fragments of DNA, one 250 base pairs in length, another 72 base pairs, somehow got inserted into the soybeans without the intent of Monsanto's genetic engineers—or had been there all along, without the company's knowledge.

The finding indicated that Monsanto scientists were not aware of the precise makeup of the genetic material they had inserted into a staple food crop. In a May 15, 2000, letter to the USDA, Monsanto acknowledged that current methods "are more sensitive and precise" than the molecular characterization techniques available in 1992, when the company previously analyzed its soybeans. The letter hastened to add that the presence of previously unknown genetic material "does not alter the initial conclusion that Roundup Ready soybean event 40-3-2 is as safe as conventional soybeans for use in food and animal feed and does not pose a risk to the environment."

Before submitting the letter to the USDA, Monsanto had undertaken its own damage control. The company enlisted an "independent panel of recognized experts" to attest to the fact that the gene-altered soybeans were as safe as ever. The USDA accepted this assessment.

Another problem with Roundup Ready soybeans had come to light in November 1999. Researchers from the University of Georgia in Athens presented a paper at the British Crop Protection Council's annual meeting in Brighton showing that the stems of Monsanto's Roundup Ready soybeans split open in extremely hot climates. Normal soybean plants, and plants engineered to resist Aventis's weedkiller, Liberty, did not experience the problem when grown in hot soil.

Bill Vencill and his colleagues had found the problem when they decided to investigate complaints made by Georgia farmers about the performance of Monsanto's gene-altered soybeans grown in conditions of heat and drought stress. They grew Roundup Ready soybeans alongside soybeans engineered to resist Liberty herbicide, glufosinate, in soils heated to various temperatures. They planted conventional soybeans as a control. In cool soils, all of the plants grew equally well. But in soil temperatures of forty-five degrees centigrade, Monsanto's soybean plants were shorter than the other varieties. The forty-five-degree soil temperature corresponded to the temperatures reached in Georgia fields when farmers suffered yield losses from Monsanto's beans, Vencill said. More troubling, the stems of nearly all of the Roundup Ready soybeans split open when the leaves began to emerge. The British magazine *New Scientist* brought widespread attention to the problem in an article by Andy Coghlan, entitled "Splitting Headache: Monsanto's Modified Soya Beans Are Cracking Up in the Heat."

Vencill and his colleagues found that the stems of Monsanto's soybeans contained 20 percent more lignin than the other varieties. Lignin makes plant tissues tough and woody. The researchers suggested that the stems of Roundup Ready soybeans split open because the extra lignin made them too brittle.

The University of Georgia study was not the first to find noteworthy differences between Roundup Ready and conventional soybean varieties. In the spring of 1999, Marc Lappé and E. Britt Bailey reported finding lower levels of phytoestrogens in Roundup Ready soybeans. Phytoestrogens are natural compounds in soy that some health experts believe provide protection against some forms of cancer. Lappé, director of the Center for Ethics and Toxics in Gualala, California, found that Roundup Ready soybeans had 12 to 14 percent less phytoestrogens than conventional soybeans grown under similar conditions.

The American Soybean Association, based in St. Louis, challenged Lappé's findings, however, posting research on its website asserting that Roundup Ready soybeans had been shown to be the same as conventional soybeans "in over 400 seed and processed fraction composition qualities, and confirmed as safe and nutritious in a series of animal feeding studies."

Curiously, Monsanto's own documents, submitted to the FDA between 1992 and 1994, mention yet another difference in composition between Roundup Ready and conventional soybeans. In a report submitted on September 2, 1994, the company said that milk produced by cows fed glyphosate-tolerant soybeans "was slightly" higher in fat than normal. The change in milk composition was noted after dairy cattle ate raw cracked soybeans for four weeks.

Despite the fact that Monsanto's soybean plants produce 20 percent more lignin in hot conditions, cause cows to make milk with a higher fat content, contain surprise fragments of foreign DNA, and possibly produce lower levels of phytoestrogens, the FDA continued to treat them as the same as conventional soybeans.

For all the official endorsements that vitamin A rice received, support for genetically engineered food continued to slide in the United States and around the world.

Food companies found themselves facing another offensive from activists in 2000: shareholder resolutions calling for a halt to gene-altered ingredients in brand-name foods. Opponents of genetically engineered foods submitted resolutions to be voted on at the annual meetings of more than twenty prominent companies, including Albertson's, Archer Daniels Midland, Bestfoods, Coca-Cola, General Mills, Heinz, Kellogg, Kroger, McDonald's, PepsiCo, Philip Morris (which owns Kraft Foods), Procter & Gamble, Quaker Oats, Safeway, and Sara Lee. Ariane van Buren, environmental director of New York's Interfaith Center on Corporate Responsibility, launched the campaign to force safety testing and labeling of genetically modified foods.

"Why rush unproven products to market? When a product is rejected by consumers or proven unsafe, shareholders are left holding the bag," van Buren wrote in an op-ed piece in USA Today in February 2000. Shareholders with over $150 million in stock "are requesting companies not to market genetically engineered food until long-term safety is assured. Meanwhile, these shareholders are asking companies to label this food to alert people who for allergic, religious or other reasons must know what they are eating."

The antibiotech resolutions that came up for a vote in the spring of 2000 garnered only slim support. At Quaker Oats Company, a proposal to stop using the new ingredients in all food products gained just 5 percent of the vote. Only 2.2 percent of shareholders at McDonald's voted in favor of such a resolution. A resolution fared slightly better at Coca-Cola, gaining 8.5 percent of the vote, but more than 98 percent of the shareholders at Philip Morris turned down a proposal to ban biotech food ingredients, as did 96 percent of shareholders at PepsiCo, 94 percent at Kellogg, and 97 percent at Safeway.

Proponents of genetically engineered food touted the votes as evidence of overwhelming public support for the continued use of the ingredients. But activists who sponsored the resolutions claimed victory in defeat. They insisted that the quick appearance of so many initiatives sent a warning to food company executives of the potential for a serious consumer backlash to materialize in the United States.

While publicly voicing support for the safety of biotech crops, some companies were quietly telling growers they would accept only conventional corn or potatoes. At the end of November 1999, McCain Foods told its growers that the company would no longer accept genetically modified potatoes for processing. Harrison McCain, chairman of the Florenceville, New Brunswick, company, told the Canadian Press that his decision was influenced by consumer pressure. "We've got too many people worried about eating the product and we're in the business of giving our customers what they want, not what we think they should have."

In February 2000 Frito-Lay announced that it would stop using gene-altered corn in its snack foods. Frito-Lay, a leader in the salty snack food industry, uses more than one billion pounds of corn to make chips and snacks each year. Frito-Lay told the Associated Press that it was responding to consumer concerns about biotech foods.

McDonald's decided in the spring of 2000 not to use Bt potatoes for its french fries. Procter & Gamble also joined the list of food companies shunning biotech ingredients, announcing that it would no longer use the gene-altered spuds in Pringles potato chips.

On April 28, 2000, responding to the decisions by restaurants and food companies to seek conventional potatoes, Monsanto issued a statement defending the safety of its NewLeaf potatoes. The company noted that consumer research "shows that American confidence in biotechnology has not changed." Nonetheless, in 2001 Monsanto withdrew its genetically engineered Bt potatoes from the market.

Greenpeace and several other activist groups launched a campaign in July 2000 aimed at pressuring Kellogg and Campbell to remove genetically engineered ingredients from the cereals and soups fed to American children. Greenpeace released correspondence showing that Kellogg was attempting to remove all such organisms from cereals fed to European children. At the same time, Kellogg was using the FDA's approval of gene-altered foods to justify leaving the ingredients in cereals fed to American youngsters.

Charles Margulis, the activist who persuaded Gerber to remove biotech ingredients from baby food, said Greenpeace had commissioned a Caravan Opinion Poll that found only 30 percent of Americans believed that Kellogg would use genetically engineered ingredients in its products. Furthermore, only 31 percent of Americans surveyed said they would buy foods from companies that promised nonengineered food to Europeans but did not make the same promise to Americans. "The real question in front of us today," Margulis said at a July 19, 2000, news briefing in Washington, "is who wants to eat this stuff?"

The Chefs Collaborative, an organization representing fifteen hundred chefs across the nation, joined U.S. activists in asking domestic producers to support the right of Americans to choose nonengineered food. "Without proper study, corporations are licensed to create organisms that have never existed before, could not have evolved without human intervention, and we know little about how they will interact with the environment," Peter Hoffman, a chef and owner of the Savoy Restaurant in New York, said at the same news briefing. "The Chefs Collaborative and all its members are committed to serving our customers, ourselves, and our children food from sources that will not only nourish our bodies but nourish the environment as well."

Hoffman joined the Chefs Collaborative in 1994, a year after it was founded. The collaborative is an offshoot of a group called Old Ways Preservation, which works to preserve traditional cuisines and cultures. Chefs in the group go directly to local farmers to buy the freshest ingredients available. "As chefs, we are trained to discern quality, flavor, and taste and are driven to explore the diversity in textures and flavors in the world of food," he said.

Knowledge about raising pure, wholesome food "needs to be on the land, in the farmer," he continued. "We pass this intelligence, this information, these products, along to our diners—mostly on the plate. People come to our restaurants knowing that we've done our homework. They entrust us not only with their appetites but with their health. They vote with their forks.

"It is critical that we maintain the ability to vote with our forks," Hoffman said. "Unfortunately, the federal government is not supporting this ability."

STARLINK AND TACOS

C lose to midnight on July 25, 2000, Larry Bohlen, an activist with the environmental group Friends of the Earth, wheeled his shopping cart to the checkout counter at a Safeway in Silver Spring, Maryland. The cart was filled to the top with food items made with corn—bags of corn chips, boxes of corn flakes, packages of taco shells and tortillas. Among the twenty-three foods he bought that night were Kraft brand Taco Bell taco shells.

His shopping spree was inspired by an item on the EPA's website about a variety of genetically engineered corn on the market called StarLink. The Bt corn was not approved for human food because it contains a substance scientists believe might cause allergic reactions in people. The EPA had approved StarLink corn only for animal feed. As Bohlen and his co-worker, Bill Freese, pored over the literature posted on StarLink corn, the words "characteristics of known allergens" popped out at them. He wondered if some of the StarLink corn meant for hogs, chickens, and cows might have made its way illegally into human food.

Before becoming an environmental activist, Bohlen had worked as a scientist for NASA; his training as a physicist had taught him the importance of testing. He shipped the corn flakes, taco shells, and other grocery store purchases to Genetic ID, the laboratory in Fairfield, Iowa, that specializes in testing foods for the presence of genetically modified organisms. He asked the company to see if any of the foods contained StarLink corn.

Genetic ID received the packages of food on August 4 and crushed them in preparation for a test using a sophisticated laboratory procedure known as a polymerase chain reaction, or PCR. The test looks for the specific DNA molecules that code for the production of a particular protein, in this case the Cry9C unique to StarLink corn.

On February 29, 2000, the EPA had assembled another Scientific Advisory Panel, including several allergy experts, to consider a request by Aventis to approve its StarLink Bt corn for human food. Very few Americans had heard of StarLink corn in early 2000. There were only a handful of reporters and members of the public in the audience at the Sheraton Crystal City Hotel in Arlington, Virginia, to hear the scientists grapple with question of whether the Cry9C protein in StarLink corn is a potential allergen. Scientific knowledge about what kinds of proteins were likely to cause allergies had not progressed much since 1994, when the EPA, the USDA, and the FDA had first discussed the topic. It was still impossible for scientists to declare with certainty that a genetically modified food would, or would not, provoke allergic responses in susceptible people.

The burden was thus on Aventis scientists to convince the EPA's scientific advisers that despite uncertainties about the Cry9C protein, StarLink corn should be approved for the human food supply. Aventis took the tack of questioning the validity of two of the four criteria that scientists had established in 1994 to screen new proteins as potential food allergens: indigestibility and stability to heat during cooking and processing. The company argued that there is no proven correlation between the stability of a protein and its potential to cause allergies. Andrew Cockburn, an Aventis official, told the EPA panel that some "less stable" proteins such as castor bean, ragweed, and fish are allergenic, while other "more stable" proteins such as horseradish peroxidase and mother's colostrum are not allergenic. Cockburn also said that a few studies had suggested that digestion might enhance the allergenicity of foods.

Cockburn contended, further, that only very small amounts of Cry9C protein would be expected to enter the human diet from StarLink corn. "The weight of scientific evidence" suggested that Cry9C would not cause allergies in people, he insisted.

EPA scientists were not swayed by Aventis's arguments. The company's line of reasoning cast doubt on the justification that federal regulators have used in approving other new gene-altered proteins for human consumption. By arguing that the stability and digestibility of a dietary protein are not necessarily good indicators of its potential to cause human allergies, Aventis was raising further scientific uncertainties about the extent to which we can know whether any biotech protein is, in fact, safe for people to eat—a fact not lost on biotechnology critics at the February 2000 EPA hearing.

Margaret Mellon of the Union of Concerned Scientists appealed to the EPA to deny the Aventis petition. "When only some of the corn is genetically engineered, and only some of that with a protein that is likely to be an allergen," she argued, it would be difficult to monitor the population for possible allergic responses to Cry9C Bt corn. "How will people and the public health officials, the pediatricians, the doctors sort that out?"

Luca Bucchini, a scientist from Environmental Defense (formerly the Environmental Defense Fund), also urged the EPA to reject the Aventis request. He argued that if the EPA were to approve the petition, it would greatly undermine consumer confidence in the agency's capacity to make a meaningful review of genetically engineered plant pesticides.

Ultimately, Aventis did not carry the day. Four months after the February deliberations, on June 29, 2000, the EPA's Scientific Advisory Panel issued its final report. "Based on the available data," the panel wrote, "there is no evidence to indicate that Cry9C is or is not a potential food allergen." The panel agreed with Aventis scientists that there is only a small amount of Cry9C protein in each kernel of StarLink corn. But the allergy experts noted that sensitive people can have severe, life-threatening reactions to even extremely low levels of an allergen. "With peanut allergies, ingestion of nanogram quantities can cause anaphylaxis," the report stated.

Thus, the question of whether Cry9C protein might cause people to have allergic reactions seemed as far from resolution in June 2000 as it had been in April 1997, when Aventis applied for a license to sell its new variety of Bt StarLink corn.

———

On August 15, 2000, a brief letter scrolled out of Larry Bohlen's fax machine from Genetic ID with the message that one of the foods he had sent to the lab, Taco Bell taco shells, had tested positive for the Cry9C protein. The box of taco shells was marked with an expiration date of September 13, 2000. Bohlen was stunned. He asked the company to repeat the test to verify that the DNA found in the taco shells could only have come from StarLink corn. "We tested the sample three times. It took a couple weeks more to test and retest a third time," Bohlen told me.

Sandy Myers, vice president for sales of Genetic ID, assured Bohlen in a September 15 letter that the testing company had a "primer set" designed specifically to detect Aventis StarLink corn. "Primer specificity has been verified by testing with this variety of corn, and with all other commercially available varieties of genetically modified corn," the letter stated.

Equipped with this verification, the Genetically Engineered Food Alert, a coalition including Friends of the Earth and six other environmental groups, went public with the information. It called a press conference on September 18, 2000, to warn American consumers that Taco Bell taco shells were adulterated with illegal genetically engineered corn. "This Cry9C is on the market illegally for human consumption. Our coalition has stepped into the vacuum left by FDA and EPA," Joseph Mendelson, an attorney with the Center for Food Safety, told reporters at the press conference in Washington, D.C.

Dan Rather led the *CBS Evening News* with a story about the contamination of taco shells with a strange variety of genetically engineered corn deemed fit only for animals. Newspapers and radio stations across the nation carried news of the StarLink corn contamination. The following day, on September 19, the Genetically Engineered Food Alert sent a letter to FDA commissioner Jane Henney asking her to order an immediate recall of Kraft Taco Bell Home Originals taco shells.

No one in the federal government had been monitoring the food supply for illegal genetic material. The FDA, which regulates twenty-five cents out of every dollar that American consumers spend, found itself in the embarrassing position of being unable to either confirm or deny the allegations of a small environmental group. In fact, neither the FDA nor Kraft had the technology in place that would allow them to test taco shells for the presence of StarLink corn. Aventis officials quickly provided the FDA with the primers and laboratory methods needed to perform such tests.

Kraft Foods, which is owned by Philip Morris Companies, posted a "special report" on its website to keep consumers abreast of the status of the as-yet-unconfirmed adulteration of its taco shells with Cry9C protein. "Our highest priority is the safety of our products. In that regard, our most immediate concern is confirming whether the protein is present in our product or not," Kraft pledged.

As press coverage intensified, dozens of people who had become ill after eating a corn product wondered if StarLink contamination might be to blame. Many people contacted the FDA or food companies to report illnesses after eating corn chips and other products made with corn.

On October 16, 2000, Grace Booth, a young woman in the San Francisco Bay area, went into anaphylactic shock after eating three enchiladas made with corn tortillas. She had not heard the news reports about StarLink corn, but two days after she returned home from the hospital, a friend mentioned the taco shell recall to her. Only then did Booth suspect that the corn tortillas in her enchiladas might have caused the severe allergic reaction she experienced.

FDA surveillance officials investigated thirty-seven illness reports filed between July 1 and November 20, 2000, which they believed were compatible with a possible allergic reaction to foods containing yellow corn flour. Many people had visited their doctor immediately, and some had been taken to hospital emergency rooms. Seventeen of those who suffered allergic reactions reported eating tacos, seven had eaten corn chips, six had consumed tortillas, and five had eaten cereal, according to Dr. Karl Klontz, a scientist with the FDA's Center for Food Safety and Applied Nutrition.

Kraft's taco shells came from Sabritas Mexicali, a wholly owned subsidiary of PepsiCo in Mexicali, Mexico. Sabritas purchases its corn flour from Azteca Milling, which processes the flour in its mill in Plainview, Texas. Azteca Milling is the U.S. leader in the production of corn masa flour for Mexican snacks, tortillas, and tacos. The company buys more than one billion pounds of corn every year.

When Azteca Milling officials read in September 18, 2000, press reports that taco shells using their flour had tested positive for genetically engineered StarLink corn, they were incredulous. "The first reaction we had was, well, that can't be, because we don't use StarLink corn. And none of our farmers use StarLink corn. None of our elevators buy it," Dan Lynn, president of Azteca Milling, told industry officials and reporters gathered in Washington, D.C., for a "StarLink Summit" sponsored by the Hudson Institute. "The way we've always done business is to use traditional varieties of corn—that's not to make a political statement for GMOs or against GMOs. Those are the varieties of corn that over time have provided us with the functional characteristics that we need in processing corn and running our business."

Azteca officials reacted swiftly to the possibility that their corn supply had been contaminated with illegal StarLink corn. Within twenty-four hours the firm stopped making yellow flour at the Plainview facility. "We sent out hundreds and hundreds and hundreds of samples of flour to two different independent testing labs—matched samples of flour—because we wanted to know what was the extent of the penetration of StarLink in our business, if indeed it had penetrated," Lynn said.

Azteca Milling spent more than $1 million in the first six weeks of testing. The test results were wildly inconsistent. "What we found was, we had corn that tested negative that resulted in flour that tested positive," said Lynn. "As we started to build a database of results from both laboratories, we found that one lab would report that a sample was positive, and the other would report that it was negative." The company decided to move its operations entirely to white corn, because there were no genetically engineered varieties of white corn on the market.

Makers of corn chips and other snacks also were blindsided by the contamination of the food supply stream with StarLink feed corn. "We did not see it coming. We didn't know it was a possibility," Elizabeth Seiler, vice president of government and regulatory affairs for the Snack Food Association, said at the StarLink Summit. "We had no idea it could happen to us."

On September 22, 2000, after tests performed by an independent laboratory confirmed the presence of StarLink corn in samples of its taco shells, Kraft Foods announced a voluntary recall of all Taco Bell Home Originals taco shell products. "Consumers who have purchased any of these products should not eat them, and should return the packages to the store where they were purchased for a full refund," the Kraft Foods website advised the public.

It was the first food product ever recalled in the United States because of a genetically engineered ingredient.

Friends of the Earth applauded Kraft's speedy voluntary recall of the adulterated products. Environmental Defense in New York also praised Kraft for doing the right thing. "No one wants unapproved pesticides in their food," said Rebecca Goldburg, senior scientist for the environmental group. "This unfortunate situation demonstrates that consumers are not being protected by the federal agencies assigned this critical responsibility. The federal government is rushing genetically engineered products to market without adequate consumer protection."

Kraft recalled more than 2.5 million boxes of taco shells from grocery stores across America. The company sells the taco shells and Mexican taco dinners under a license from the fast-food chain Taco Bell. Sales from the taco line of foods account for about $50 million of Kraft's total annual revenue of $27 billion, according to *The Wall Street Journal*.

Betsy Holden, president and CEO of Kraft Foods, called for stronger regulation of biotechnology companies. To prevent this kind of contamination from ever happening again, Holden said, the federal government should discontinue the practice of approving a food crop for use as animal feed only. There should be mandatory reviews of all biotech plants before they enter the market. And as a precondition of approval, regulatory agencies should require biotech companies to have a validated testing procedure ready for identifying the DNA in gene-altered crops and finished food products.

The Grocery Manufacturers of America issued a statement applauding Kraft Foods for acting in the best interests of its customers. GMA had long been an ardent booster of biotechnology, yet the trade group now endorsed tougher federal oversight. "We believe that this situation suggests there may be some improvements needed in the regulatory system," the GMA said. "Kraft has made specific suggestions for enhancing the safe entry of biotech foods into the marketplace that we believe merit serious consideration."

With the FDA still unable to confirm whether the taco shells contained illegal corn, the National Corn Growers Association backed Kraft's voluntary recall. "At stake, is the reputation and integrity of U.S. corn farmers and the food they produce," said Roger Pine, head of the 32,000-member association. If confirmed, the presence of StarLink DNA in the tacos "would indicate a significant breach" in the Aventis license, Pine charged. "Until that question can be answered conclusively, we support Kraft's voluntary withdrawal of the taco shells."

Widespread news reports about the contamination of Taco Bell's shells left American consumers bewildered about what StarLink corn was and where it had come from. Many people were so worried the genetically engineered variety would sicken their families that they stopped buying any processed foods listing corn as an ingredient.

Congressional backers of genetically modified foods were also thrown off guard. Republican senator Tim Hutchinson of Arkansas and Democratic senator Christopher Dodd of Connecticut formed a congressional Biotechnology Caucus to investigate federal regulation of the foods. Jim Jeffords of Vermont hurriedly put together a Senate hearing on the erosion of public confidence in biotechnology and its future in the U.S. food supply.

"The challenge here is to seek out the appropriate balance between promising technological innovation and prudence," Senator Jeffords said, opening the oversight hearing on September 26, 2000. "Without that balance, nothing we do here in Congress, in academia, in industry, or on the farms will make a whit of difference, if American consumers lack confidence about the safety of the foods they are eating."

The director of the FDA's Center for Food Safety and Applied Nutrition, Joe Levitt, defended the agency's handling of genetically modified foods. "FDA has found no evidence to indicate that the DNA inserted into plants using bioengineering presents any safety problems," Levitt told the senators. He denied the charge that biotech crops were getting "a free ride" through the regulatory system.

Tom Harkin, whose Iowa constituents grow much of the nation's corn, voiced his belief that biotechnology would bring great benefits to society, but at the same time he worried aloud that the science has raced far ahead of consumer information about genetically engineered food. The Democratic senator was shocked and angered that it was an environmental group, Friends of the Earth, and not one of the three federal agencies charged with protecting public health and safety, that had found the illegal corn. "FDA lacks the wherewithal" to effectively oversee the biotech food industry, he charged.

Senator Barbara Boxer made a plea to her colleagues to support her labeling bill. In February 2000 the California Democrat had introduced a bill in the Senate requiring that all genetically engineered foods be labeled. "The health and environmental effects of genetically engineered food are not yet known," she wrote in a press statement introducing the bill. Boxer's legislation was a companion to the labeling bill that Representative Dennis Kucinich had introduced in the House in November 1999.

"The bill takes no position on whether genetically engineered foods are safe or whether they are harmful," Boxer said. "The bill authorizes long-term studies on the safety of these foods. Such studies simply do not exist today."

Boxer told her colleagues at the hearing that a recent public opinion poll had found that an astounding 92 percent of consumers favored mandatory labels on gene-altered foods, yet the Grocery Manufacturers of America continued to fight such labeling.

"Given the great public and international support for labeling, the amount of controversy generated by my labeling bill is almost unbelievable," Boxer added. "In a culture where we place a high value on the rights of individuals to make decisions based upon accurate information, opposition to labeling is out of step."

While Congress was investigating the lapse in FDA and EPA oversight of gene-altered crops, food makers were scrambling to devise ways to keep the illegal corn out of their products. Reuters reported on September 26, 2000, that Quaker Oats Company would require suppliers to provide genetic identification for the grains it purchased.

The company specified that StarLink corn would not be accepted. The division of ConAgra that supplied taco shells to Taco Bell restaurants began purchasing corn flour from a mill in Indiana instead of from its regular source in Texas.

Also on September 26 Aventis CropScience suspended sales of Star-Link corn hybrids for the 2001 growing season. Aventis CropScience, located in Research Triangle Park, North Carolina, is the agricultural biotechnology division of Aventis responsible for stewardship of Star-Link corn. Three days later, at the urging of the USDA and the EPA, Aventis announced it would buy back from farmers as much of the 2000 harvest of StarLink corn as could be located. The company reached an agreement with the USDA under which the government would purchase the 2000 StarLink harvest, plus all "buffer corn"— regular, non-StarLink corn grown within 660 feet of StarLink—unless farmers planned to keep the corn on their farms to use as cow, chicken, and hog feed. The USDA stepped in to administer details of the buy-back program.

Under the unprecedented scheme, the USDA offered to pay growers the October 2, 2000, posted county price, plus a twenty-five-cent pre-mium per bushel, for the StarLink corn and buffer-zone corn, which was likely to be contaminated with StarLink genes through cross-pollination. Aventis pledged to repay the government for the cost of the program.

Even as Aventis and the USDA vowed to round up the entire 2000 StarLink grain harvest and channel it away from the food chain, it be-came clear that the contamination of Taco Bell shells was not an iso-lated incident. In mid-October Safeway revealed that its house brand taco shells too had tested positive for StarLink corn. The national chain said it would immediately remove all Safeway-brand taco shells from store shelves. Officials from Mission Foods in Texas, which makes Safeway taco shells, said the corn flour had come from Azteca Milling, the same mill that supplied the flour for Kraft's Taco Bell taco shells.

As Aventis and the USDA began tracking down farmers who had purchased StarLink seed corn, they learned that some growers had no idea they were supposed to have sold the grain exclusively for animal feed or industrial uses. Some farmers said their seed dealers had never told them of the restrictions. Others said they had been aware of the restrictions but not of the requirement for a wide buffer zone. Many growers insisted they never would have bought, or planted, StarLink corn had they known about the restrictions. Ralph Klemme, an Iowa legislator, told a Reuters reporter that he had not been fully informed about the restrictions when he planted twenty-five acres of StarLink corn. Klemme, who chairs the Iowa House agriculture committee, said he would not have planted StarLink corn had he known that other varieties of corn were not to be grown within a wide buffer zone, to avoid cross-pollination.

On October 12, 2000, Aventis voluntarily asked the EPA to cancel its license to sell StarLink corn in the United States.

The extent of adulteration of the food supply became apparent on November 1, 2000, when the FDA issued a Class II recall of three hundred products contaminated with StarLink corn. The recall covered tortillas, taco shells, tostadas, and chips manufactured by Mission Foods. A Class II recall means that people eating the food may suffer "temporary or medically reversible adverse health consequences." Two weeks later, on November 15, the FDA issued a Class II recall for cornmeal, corn flour, snack meal, and flaking grits contaminated by StarLink corn, affecting dozens of products manufactured by the ConAgra Grain Processing Company of Omaha, Nebraska.

Food analysts quickly realized that the StarLink contamination of processed foods probably was not confined to the 2000 corn harvest. The September 2000 expiration date on the box of contaminated Taco Bell taco shells meant that the tainted Cry9C corn had probably been harvested in 1999. Farmers and food company officials agreed it was highly unlikely that corn grown in the summer of 2000 could have made its way through the grain handling system and into a finished box of shells in so short a time.

StarLink represented only a tiny fraction of the U.S. corn acreage in the three years it was planted. In 1998, according to the EPA, farmers planted StarLink seed on only 9,018 acres out of a total of 80 million acres of corn. In 1999 it was planted on 247,694 acres and in 2000 on 350,420. Aventis insisted it would be able to capture most of the 2000 StarLink crop before growers moved the harvested grain off their farms. The problem facing U.S. grain handlers and food manufacturers, though, could be summed up in one word: commingling. A single truckload of StarLink corn could contaminate all the corn in a grain elevator. There is no way to distinguish StarLink from other corn by its physical appearance—it looks just like all the other Number 2 yellow corn moving through the grain system. Once shipped to millers and food processors, StarLink was mixed with normal food corn. Some industry officials estimated that even though StarLink made up little more than 1 percent of Iowa's corn harvest in 2000, as much as half of the state's corn could have become contaminated through mixing and dilution.

For the food industry, the prospect of any detectable quantities of illegal—and potentially allergenic—corn getting into all manner of trusted, brand-name products was catastrophic. Across the farm belt, grain elevator operators had started testing each truckload for contamination by StarLink corn. Food companies were testing corn flour. The Associated Press reported that in mid-October, when testing delayed deliveries of grain, a major breakfast cereal company temporarily shut down a production line. By December 9, 2000, the AP reported, supplies of Cheetos, made by Frito-Lay, were down by as much as 10 percent because the company was running "thousands of tests a month to make sure" its cornmeal supplies did not contain any StarLink corn.

DNA testing is expensive. According to Susan Harlander, a food industry consultant and former vice president of Pillsbury, costs vary from $150 to $450 per test. "Current methodology has a high rate of false positives and false negatives. Additionally, tests must be customized for each product category, since matrix effects are significant," Harlander told the EPA. The matrix is the final form the processed corn takes in a food product. According to Aventis officials, more of the Cry9C protein could be expected to survive in the form of cornbread, hush puppies, and polenta, which are made from cornmeal, than in tortillas or tacos. This is because corn for tortillas is cooked and steeped in lime, which leaches out much of the protein, before it is formed into dough, baked, and fried. EPA scientists believe there is virtually no protein in high-fructose corn syrup, corn oil, or cornstarch, all of which go through a process of "wet" milling.

Faced with the expense of repeatedly testing Number 2 yellow corn—and not knowing for certain when or if a minute quantity of StarLink might slip into their products—some food manufacturers switched to white corn. White corn costs about 10 percent more than yellow corn. The only way a food maker that used yellow corn flour could be absolutely certain its supply was free from the Cry9C gene was to import the grain from a country where genetically engineered corn was not grown.

If the domestic grain handling, transportation, food manufacturing, and retail sectors were thrown into chaos, many Asian buyers of U.S. corn were seized by panic. American farmers export between 50 and 55 million tons of corn each year. Japan, the largest buyer, typically purchases between 15 and 16 million tons. Korea is second, with imports of about 8 million tons. Japanese consumers, like Europeans, have soundly rejected genetically modified foods. Japan had recently passed a law, due to take effect on April 1, 2001, setting a zero tolerance for the import of unapproved agricultural products. StarLink corn was approved neither for human food nor for livestock feed in Japan, so it was essential that not a trace of the corn make its way into the country.

Within a week of the Friends of the Earth announcement that U.S. tacos were adulterated, USDA officials moved to shore up confidence among Asian buyers. On September 25, 2000, Isi Siddiqui, special assistant for trade to the secretary of agriculture, led a trade mission to Japan to calm public fears. USDA officials and representatives of the U.S. Grains Council and the National Corn Growers Association spent two days in intense meetings with the Japan Corn Foods Association, the Japan Feed Trade Association, Zennoh, the largest co-op in Japan, the Japan Ministry of Health and Welfare, and the Ministry of Agriculture, Forestry, and Fisheries.

USDA officials negotiated an agreement requiring all American corn intended for food production to be tested before being unloaded on Japanese docks. About 4 million tons of the annual Japanese corn imports from the United States are for food, while the remaining 11 to 12 million tons are used for animal feed.

Despite U.S. government assurances and widespread DNA testing of corn shipments, however, StarLink corn did find its way into food products sold in Japan. On October 26, 2000, Andrew Pollack of *The New York Times* reported that a Japanese consumer group, the No GMO Campaign, had detected StarLink corn in a cornmeal product called Homemade Baking. The Japanese government asked the cornmeal manufacturer, Kyoritsu Shokuhin, to recall the product until health officials could perform independent tests to determine whether the foods were contaminated by StarLink. The following day the USDA issued a guidance to exporters reminding them of the prohibition against shipping StarLink corn to Japan. The USDA also sent a scientist to Tokyo to quell Japanese worries about the seeming inability of the United States to keep StarLink out of their nation's food supply.

In November 2000 Kyodo News Service reported that the Ministry of Health and Welfare had confirmed the presence of the banned StarLink corn in food products sold by retailers in Japan. By then the USDA had acknowledged a steep drop-off in American corn sales to Japan. Philip Brasher, Associated Press farm writer, reported on November 16, 2000, that during the week ending November 9, Japan had purchased less than half its normal amount of U.S. corn. "Agriculture Secretary Dan Glickman acknowledged that StarLink is probably a cause of the export drop," Brasher wrote. Traces of StarLink corn continued to turn up in Japanese food products in January and February 2001.

EPA officials estimated it would take four years for the StarLink corn to entirely work its way out of the U.S. food chain—a calculation based on the optimistic assumption that no corn seed tainted with the Cry9C gene would ever be planted again.

In the fall of 2000, as the StarLink crisis deepened at home and abroad, Aventis again asked the EPA to approve the corn variety for human food. At the end of October, the company submitted a petition seeking a four-year exemption from the prohibition of StarLink in human food. Aventis, like the EPA, speculated that within four years, all of the StarLink corn already in the grain stream would work its way out of the U.S. food chain. The company argued that new scientific evidence suggested that StarLink corn is as safe for human consumption as other Bt corn varieties. The petition enjoyed the strong backing of food industry trade groups and biotechnology proponents, who argued that the exemption was necessary to avert major disruptions in U.S. food production.

Aventis claimed that a new study had shown that the Cry9C protein remained stable for only thirty minutes in simulated gastric juices at a pH of 1.2. In previous studies performed at pH 2.0, the Cry9C protein had been stable for more than four hours. Since stability to digestion is one of the accepted attributes of known allergens—and the new studies showed Cry9C protein breaks down under some conditions—Aventis argued that the allergy concern was diminished.

The EPA's scientists, however, were not so sure about the value of the new study. In a review of the Aventis petition, the agency pointed out that scientists had recently learned that "normal" stomach acidity can vary greatly from one person to another. In fact, the gastric pH of normal individuals who have not eaten food in several hours can vary from less than 1.0 to 3.0—a hundred-fold difference (pH increases logarithmically) in acidity.

"This pH can be raised significantly, but temporarily, by ingestion of food, antacids, or medications," EPA scientists said in a report. Further, the transit time of food through the stomach can vary greatly from one individual to another. The average time for digestion is one hour, but food can pass through the stomach in as little as fifteen minutes when a person has eaten just a small item on an empty stomach, or it can take as long as four hours after a full meal.

Aventis also argued that the chance of Americans having been exposed to much Cry9C protein was low. This was important because before suffering an allergic reaction to a protein, a person must first become sensitized to it. Sensitization, in turn, requires exposure to the offending protein. Aventis contended that the U.S. population was unlikely to have been exposed to the Cry9C protein from sources other than StarLink corn. The Bt subspecies *tolworthi* bacterium was found in the Philippines and is not known to occur in the United States. In addition, the Cry9C protein represented only a tiny fraction of the total protein content of a each kernel of StarLink corn. These factors, Aventis argued, suggested that there was a only minimal potential for Americans to become sensitized to the Cry9C protein.

Aventis reached this conclusion by estimating the amount of Star-

Link corn a person would have to consume to become sensitized to the Cry9C protein. As little as 16 micrograms of a particular protein in peanuts can elicit an allergic reaction in sensitized people. Aventis performed a dietary analysis, which was based on the finite amount of StarLink-brand corn known to have been grown in the United States. According to the company's calculations, people who ate the maximum possible amount of corn would take in only 6.4 to 8.6 micrograms of Cry9C protein—well below the 16-microgram threshold needed to elicit peanut allergy reactions.

But EPA scientists questioned the validity of making a dietary comparison between peanut allergen and Cry9C protein. When it performed its own dietary analysis, the agency came up with several scenarios in which children, in particular, could approach or even exceed the threshold of 16 micrograms for the intake of Cry9C protein.

For the EPA, the StarLink issue was fraught with political as well as scientific controversy. Repercussions for U.S. consumers, food companies, and corn growers were growing with each passing week. EPA regulators said they believed that Americans would be exposed to extremely low amounts of Cry9C protein from eating foods containing StarLink corn, but they nonetheless took a cautious approach to the Aventis petition. Responding to a situation many in the food industry described as "intolerable," the EPA scheduled an urgent meeting of its science advisers for November 28, 2000.

This Scientific Advisory Panel of world-renowned allergy experts would decide whether exposure to StarLink corn posed "an unacceptable risk" for the U.S. population. Some of the experts chosen for the panel had served on the panel convened by the EPA in February 2000 to assess the potential for Cry9C protein to cause allergies. The EPA also invited members of the public to express their views on whether the government should approve the Aventis request to legalize StarLink in human food for four years.

Just a week shy of the EPA's slated Scientific Advisory Panel meeting, Aventis stunned the food industry with yet another disturbing revelation: Cry9C had been found in a variety of corn seed that had no known connection to StarLink corn. "Aventis does not know how Cry9C protein came to be present in a variety other than StarLink-brand seeds," the company said on November 21, 2000.

Aventis had apparently discovered its patented gene in another variety of corn after several farmers had been turned away from grain elevators because their grain was contaminated with StarLink. The farmers insisted they had never purchased StarLink seed, yet DNA tests on their harvested corn were positive for Cry9C. The new corn containing the Cry9C gene was traced to Garst Seed Company in Slater, Iowa. Garst Seed was the company licensed by Aventis to make StarLink Bt corn.

Garst officials said they discovered the Cry9C protein in a non-StarLink corn hybrid that the company had produced in 1998. Garst Seed president David Witherspoon said he did not know how the Cry9C protein got there.

The contamination might have resulted from the mishandling or mislabeling of bags of StarLink and non-StarLink corn seed. Alternatively, pollen from StarLink might have blown over nearby cornfields and cross-pollinated with the non-StarLink corn, thus moving the Cry9C gene into a nonengineered Garst hybrid.

No one familiar with the grain industry was truly surprised that genes from one biotech variety of corn had made their way into another corn hybrid. Beginning with the first biotech harvest in the fall of 1996, both the USDA and Monsanto had argued that the U.S. grain handling system was not equipped to segregate conventional corn and soybeans from gene-altered ones, particularly when there were no distinguishing physical characteristics between the varieties. In the wake of the StarLink incident, USDA officials and biotech companies that had vehemently resisted European calls for crop segregation found themselves scrambling to ferret out minute caches of illegal StarLink corn commingled in the vast streams of grain moving on conveyor belts, trucks, trains, and barges from farms to food processing plants across America.

Aventis's justification for asking the EPA to retroactively approve Cry9C protein rested on the premise that the amount of StarLink corn in existence was both small and finite. Careful calculations had been made to determine worst-case scenarios for human exposures to the corn. But with the Cry9C protein turning up in a rogue variety, no one could say for certain how much Cry9C protein might find its way into the food supply.

———

Vigorous press coverage of the StarLink contamination of tacos and corn chips was beginning to take a toll on public confidence in genetically engineered foods. In a November 6, 2000, commentary, *Business Week* slammed the biotech industry for its continued "unyielding opposition" to mandatory labeling and special regulations. "Biotech foods are new, they are different, and they deserve special regulations," reporters for the respected business magazine wrote. "The industry should drop its opposition to tougher regulations. That could boost consumer confidence and disarm the critics. Then we might all begin to enjoy, much sooner, the benefits that biotech foods can provide."

Food industry trade groups, instead of following this advice, dug in to fight renewed calls for labels. Biotechnology proponents had already concluded that StarLink corn was totally harmless, but in the ongoing news reports of food product recalls, that perspective was nowhere to be found. On the eve of the EPA's high-profile Scientific Advisory Panel meeting, two of the biotech food industry's prominent opinion-makers published op-ed pieces in major daily newspapers downplaying the significance of StarLink Bt corn.

"Nobody has gotten an allergic reaction from the corn. The evidence says convincingly that no one will," Dennis Avery of the Hudson Institute in Indianapolis wrote in the November 26, 2000, *Chicago Tribune*. "Independent laboratories have found nothing to confirm the EPA's original suggestion that StarLink corn might be allergenic. But the EPA's blunder has helped poison the well for biotech crops."

Cry9C protein makes up just over one-tenth of one percent of the total proteins in StarLink corn, Avery pointed out. "A single serving of peanut butter doses the consumer with 1.3 grams of peanut allergen," he noted. "That's more than fifty thousand times the Cry9C protein even the most ardent taco lover would ingest in a year!"

The Washington Post ran an illustrated half-page article on November 26, 2000, entitled "Tacogate: There Is Barely a Kernel of Truth." The opinion piece portrayed the StarLink contamination episode as a trifle. "For all its ominous overtones, the StarLink incident has very little to do with science and safety," wrote Thomas Hoban, a professor of food science and sociology at North Carolina State University. StarLink corn "is really no different from other corn, except for the addition of a gene that produces an insect-fighting protein." Hoban quoted another scientist, Steve Taylor, head of the University of Nebraska's department of food science and technology, as saying "there is virtually no risk associated with the ingestion of StarLink corn in this situation."

The article glossed over Aventis's failure to keep StarLink corn out of the human food chain. Instead, Hoban shifted blame from Aventis to the EPA and the environmental groups that had blown the whistle on the tainted corn. It was the EPA and the antibiotechnology activist groups that "need[ed] to act responsibly," he insisted. "It's not reasonable to demand 'zero risk' from any technology, nor to hold biotechnology to unreasonably high standards."

Hoban also warned against mandatory labeling, voicing the habitual industry argument that labels would serve only to "confuse" consumers and make food more expensive. "Opponents who call for mandatory labeling of all foods with biotech ingredients do so mainly as a means of launching a further attack on the industry. The FDA already requires nutritional and health labeling, and research has shown that a simple statement that a food 'contains genetically modified ingredients' would serve chiefly to confuse and alarm consumers."

In any case, Hoban concluded, American consumers had more important things to worry about than the illegal release of a potential new allergen into the food supply. "Most of us have enough daily concerns without being frightened into thinking the food we're eating is dangerous," he wrote.

Thomas Hoban, who has close ties with the food industry, frequently advises both government officials and industry trade groups on public attitudes toward genetically modified foods. He wrote one of the State Department's WTO briefing papers on agricultural biotechnology trade issues. He is a frequent speaker at food industry meetings and also has served as an adjunct faculty member at the Georgetown University Center for Food and Nutrition Policy in Washington, D.C.

The Center for Food and Nutrition Policy, which switched its affiliation in September 2001 from Georgetown University to Virginia Polytechnic Institute and State University in Blacksburg, Virginia, describes itself as "a non-partisan group dedicated to the ideals of advancing food safety, biotechnology and nutrition in America and around the world through science." But the center is actually entangled with both the biotechnology industry and the Grocery Manufacturers of America, through linked board members and cooperative funding arrangements. Lester Crawford, who became director of the center in July 1997, formerly served as executive vice president of the National Food Processors Association. From 1987 to 1991, Crawford was administrator of the USDA's Food Safety and Inspection Service, where he was charged with oversight of the biotech food industry. He also previously served as director of the FDA's Center for Veterinary Medicine, in which capacity he was involved in the review process for approval of Monsanto's genetically engineered bovine growth hormone, rBGH.

In November 1998 the Georgetown University Center for Food and Nutrition Policy entered into a "strategic alliance" with the Grocery Manufacturers of America. Manly Molpus, GMA's chief executive officer, served on the center's board of advisers in 2000. Frank Kotsonis of Monsanto and Klaus von Grebmer of Novartis were also members of the center's board in 2000.

Since the early 1990s, Hoban had been conducting surveys for the food industry on public attitudes toward biotech foods. In the wake of media stories linking gene-altered corn to taco shell recalls, other national surveys found a wary citizenry, but Hoban found Americans high on gene-altered foods. On October 12, 2000, GMA released a poll designed and analyzed by Hoban showing that "Americans remain positive over the benefits of agricultural biotechnology."

Two-thirds of Americans surveyed in Hoban's poll said they "would be likely to buy produce such as potatoes or tomatoes that had been modified through biotechnology to require fewer pesticides." Two-thirds also said they would buy produce modified to contain more vitamins and nutrients. "This is basically the same response we have seen over the past five years to the same question on other national polls," Hoban said of the October 2000 survey results.

Polls conducted by other organizations in the fall of 2000, however, suggested Americans were growing nervous about genetically engineered foods. A Reuters/Zogby poll released on November 3, 2000, just three weeks after Hoban's survey for the GMA, found that a majority of Americans believed the StarLink corn recalls of foods containing genetically engineered corn raised concerns about the safety of the U.S. food supply. Nearly 60 percent of women expressed concern about the recalls. Only 25 percent of respondents said they were not concerned.

One-third of the 1,210 people surveyed in the Reuters/Zogby poll said farmers should not be allowed to grow gene-altered crops. Nearly 40 percent of respondents said farmers should be allowed to plant gene-altered crops, while 20 percent said they were not sure.

Yet another national poll conducted in this time period, an Oxygen/Markle Pulse Foundation survey taken on November 11, 2000, revealed that many Americans were concerned about the risks to themselves and their families from eating gene-altered foods. In the Pulse survey, only 50 percent of American women said they would eat genetically modified foods, while only 37 percent said they would feed those foods to their children. The poll found that only 38 percent of women and 50 percent of men supported the genetic modification of food. Nearly half of the women surveyed said they were willing to pay more for nonengineered food.

An overwhelming majority—88 percent of respondents in the Oxygen/Markle Pulse survey—supported safety testing; 85 percent supported labeling. Most Americans judged themselves to be uninformed about genetically modified foods. Only 18 percent responded that they felt "very informed" about the benefits and risks of such foods.

The contamination of hundreds of products with illegal StarLink corn strained the ability of the biotech food lobby to neutralize the public's growing uneasiness about genetically engineered food. Some food industry analysts began to question openly why name-brand food companies should keep "taking a bullet" for agricultural biotechnology companies. It was, after all, Kraft, Campbell, and Kellogg whose reputations were tarnished when gene-altered food crops went wrong—not Aventis, Monsanto, and Novartis, the companies that produced and profited from the seeds.

Promar International, an Alexandria, Virginia–based consulting firm, warned food companies that they needed to come up with a strategy for handling genetically modified ingredients in a post-StarLink marketplace. "Many observers fear that this could be the event the food industry has dreaded: large numbers of consumers have finally focused on GMO ingredients and what they see frightens them," Promar analysts said.

Executives in food, beverage, and agriculture companies "should be thinking about how they will deal with the inevitable next case," Promar warned in a report. The consultants urged food industry executives to consider their exposure to gene-altered foods.

On November 27, 2000, the day before the highly anticipated EPA science meeting on the allergenicity of StarLink corn, the GMA and the Alliance for Better Foods held a luncheon briefing for reporters at the Willard Hotel in Washington, D.C. Lisa Katic, a GMA official and a representative of the alliance, said industry scientists "believe that the science-based safety assessment demonstrates that the potential exposure to StarLink is very low and that the risk of allergenic reactions is remote."

Lester Crawford of the Georgetown University Center for Food and Nutrition Policy was among the experts invited to the briefing. When I asked him what kind of impact he thought the StarLink corn contamination would have on the future acceptance of gene-altered foods in America and abroad, his response was upbeat. "The StarLink incident hasn't changed public opinion very much at all in the U.S.," he said. "In the rest of the world, it couldn't be worse. But you know, when they see things like this golden rice and nonallergenic peanuts, they'll come around."

The day after the luncheon, in a crowded hotel meeting room in Rosslyn, Virginia, the EPA's Scientific Advisory Panel once again took up the question of whether StarLink corn might cause allergies. The EPA asked the panel of thirteen distinguished allergists and plant science experts to decide whether the Cry9C protein in StarLink corn had a high, medium, or low risk of causing allergies. Federal officials opened the public meeting by describing the scope of the government's efforts to contain and deal with the StarLink crisis.

An official with the Centers for Disease Control and Prevention (CDC) in Atlanta first discussed the ongoing investigation into the public health effects of StarLink in the U.S. food supply. On November 8, 2000, the FDA had asked the CDC to open a formal, rapid-response epidemiological investigation into the cases of people who might have experienced allergic reactions to StarLink corn.

The investigation was based entirely on "passive reporting" by people who had voluntarily told the FDA or the EPA that they had become ill after eating a corn product. Neither CDC nor FDA officials had made any attempt to look for people who might have experienced reactions to StarLink corn without realizing the source of their illness. Federal agencies did not alert allergists to watch for possible reactions to corn products. Most of the people who called the FDA did so after hearing news reports following the September 18, 2000, revelation that taco shells were being recalled because of StarLink corn contamination. The FDA did refer to CDC investigators two people who had reported becoming ill after eating corn products in July and August, before the widespread media reporting on StarLink.

The CDC contacted several dozen people who had reported becoming ill and asked them to fill out a lengthy health questionnaire. Nearly everyone cooperated, Carol Rubin, a member of CDC's investigation team, told the EPA advisory panel. The cases of possible allergic reactions to StarLink corn were scattered across the country—from Washington state to Florida, and from Georgia to Wisconsin. Investigators took blood samples from seventeen people whose symptoms were most suggestive of an allergic response. But the CDC investigation hit a roadblock in November 2000: no laboratory test existed that could determine whether people were allergic to the new genetically engineered protein.

Even with the cooperation of patients and state health departments, this type of investigation was fairly limited, Carol Rubin told the EPA science advisers: "No matter how thoroughly we review these adverse event reports . . . we still will not be able to determine if the reported symptoms were due to consumption of products containing StarLink corn or if any individual case has antibodies against, or is allergic to, the corn product."

To solve the StarLink corn allergy puzzle, the government would have to develop a reliable laboratory method to test human blood for antibodies against the corn.

After FDA and CDC officials described their investigation, the scientific panel heard the views of thirty-six members of the public, who were split equally between representatives of industry groups and consumer organizations. Grain handlers and food processors supported the Aventis petition to allow StarLink corn in the food supply for four years. Many made a plea to the government for pragmatism. They argued that it was not practical for the government to declare foods containing even minuscule amounts of StarLink corn as adulterated and subject to recall.

Michael Phillips of the Biotechnology Industry Organization, called the situation confronting the food industry intolerable. "Currently, if one kernel or less of StarLink corn is detected, either in a shipment or in a product, it must be considered adulterated. In other words, we're dealing with a zero tolerance," said Phillips. "We deal with that very rarely within our total food system."

Consumer groups, meanwhile, urged the EPA to reject the Aventis petition. "To grant the exemption would erode public trust in the safety and integrity of the United States food supply," argued Margaret Wittenberg, a representative of Whole Foods Market. The Austin, Texas–based company runs 118 supermarkets nationwide that specialize in organic and natural foods. Wittenberg told the EPA that customers were understandably upset when they learned StarLink corn was in processed foods. Because the company could not guarantee that products processed from yellow corn from the U.S. commodity market were free of the gene-altered variety, many customers had stopped buying foods made with yellow corn altogether.

"The fact that the Cry9C protein that was supposed to be exclusive to Aventis StarLink corn was also found in another variety of corn illustrates that gene drift, a probable cause of the contamination, is a very real problem," Wittenberg said, adding, "Its discovery in another variety may be just the tip of the iceberg."

Larry Bohlen, who first discovered StarLink corn in taco shells, pleaded with members of the EPA science panel to consider the victims who suspect they might have developed a serious allergy to Cry9C. He recounted the harrowing experience of Sharon Wolff, whose young son had trouble breathing after dinner on the evening of July 3, 2000. The four-year-old was rushed by ambulance to the emergency room of

Alexian Brothers Medical Center in Elk Grove Village, Illinois, where doctors gave him oxygen. Emergency room physicians determined that the child was suffering a severe allergic reaction to a food he had ingested within the previous two hours. Among the foods the boy ate for dinner were packaged frozen chicken breast nuggets coated with corn flour, honey, and chocolate graham crackers. The doctors tested him for allergies to chicken, chocolate, and honey. All the tests came back negative.

"Due to the mysterious and sudden nature of my son's reaction, the allergist requires us to now carry a Pediatric Epipen and Benadryl with us at all times in case this should recur," his mother wrote in a November 25, 2000, letter to EPA administrator Carol Browner. The boy is not allowed to eat chicken or chocolate outside of his home— although he has eaten those foods at home subsequently with no problem. Over the next few weeks, the Wolffs remained mystified as to what could have caused their son's allergic reaction. After hearing news reports about StarLink corn in the fall, Sharon Wolff read the list of ingredients on the box of chicken nuggets and saw that it included yellow corn flour.

In her letter, Wolff asked EPA to "please consider what unnatural ingredients can do to a small little trusting body, such as my son's. Would ingredients be acceptable for adult size consumers in the short run, and dangerous or questionable for children?"

As the day-long meeting stretched into night, the physicians on the panel made it clear that they harbored serious doubts about the wisdom of approving the entry of even limited amounts of StarLink corn into the human food supply. Several of the doctors on the November 28, 2000, panel who had also served on EPA's panel in February 2000 pointed out that there had been no new breakthroughs in research about allergies—or the Cry9C protein—in the intervening eight months.

Hugh Sampson, a pediatrician at New York University Medical Center, voiced worries about the possibility that children with allergies might become exposed to higher levels of StarLink corn in their diets than the general public. Corn allergy is extremely rare, which means that children who are allergic to milk, wheat, and other foods end up eating a lot of corn, he explained. "What concerns me as an individual who treats a lot of highly allergic children is that corn is the food that we use as our major food source." Aventis officials did not consider the diets of such high-risk children when it calculated that Americans would have low exposure to StarLink corn.

Some of the children are ingesting corn in a variety of ways—tacos, corn chips, grits—"so these children who are at highest risk for developing allergies are getting way over the levels they were predicting for the corn," Dr. Sampson said. "And if this is a potential allergen, that would be of great concern."

In the end, the Scientific Advisory Panel agreed there was a "medium likelihood" that the Cry9C protein in StarLink was a potential allergen.

"I would err on the side of caution," Marc Rothenberg, an allergy and immunology specialist at Children's Hospital Medical Center in Cincinnati, Ohio, told EPA officials. "I haven't seen any convincing scientific evidence disproving this is an allergen."

In the months following the November 2000 meeting, FDA scientists set about creating a laboratory test for the Cry9C protein. The agency used proteins provided by Aventis to develop a test that could detect immunoglobulin E, or IgE antibodies, in human blood. IgE allergic reactions affect between 10 and 25 percent of the population in most developed countries, according to the World Health Organization. People can have IgE allergic reactions to pollen, mold spores, animal dander, insect venom, and other agents besides food. IgE allergic reactions take place within a few minutes to a few hours after ingesting the offending food. Between 5 and 8 percent of children under the age of three suffer from IgE allergies to a particular food. Among adults the incidence is lower, estimated at between 1 and 2 percent.

To have an IgE reaction, a person must first be exposed to a protein that causes the body to produce the antibodies. The antibodies attach to cells in the body, which then circulate in the blood. The next time the sensitized person eats the problem food, the antibodies trigger an allergic reaction—which can range from mild to severe to fatal.

In March 2001 the FDA completed work on an assay to test human blood for antibodies to the Cry9C protein. CDC officials sent blood samples taken from seventeen of the potential StarLink corn allergy victims to the FDA to begin testing. In June the CDC issued a report on its eight-month investigation: it concluded that none of the seventeen human blood samples tested using a new "ELISA" assay developed by the FDA showed antibodies to the Cry9C protein.

Aventis again petitioned the EPA, this time asking the agency to grant a tolerance, or maximum allowable level, of 20 parts per billion for the presence of StarLink corn in U.S. food products. To consider the new Aventis request, the EPA called another Scientific Advisory Panel meeting on July 17–18, 2001.

"Today, we are in a very different place than we were back in November," Janet Andersen of EPA's Biopesticides Division said at the opening of the July meeting. The EPA had revised downward its estimate of the amount of Cry9C pesticide residues that were likely to contaminate U.S. food corn. It calculated that the average residues most likely ranged from less than 1 part per billion to 8 parts per billion—significantly less than the 20 parts per billion Aventis was asking the EPA to allow in food.

FDA and CDC officials presented findings to the advisory panel suggesting there was no evidence that the Cry9C protein causes allergies. Several members of the advisory panel sharply questioned the validity of the FDA's new laboratory test, however. These allergists were troubled by the fact that the test was based on the Cry9C Bt protein produced in *E. coli* bacteria, rather than on the actual Cry9C Bt protein produced in StarLink corn plants.

Dean Metcalfe, an allergy specialist at the National Institutes of Health, argued that small differences between the bacterial and plant forms of Cry9C cast doubt on the certainty of the FDA's findings. While the FDA's test gave a relative reassurance that an individual did not have antibodies to Cry9C, Metcalfe said, it was not absolute.

Marc Rothenberg, another member of the advisory panel who had served on the November 2000 panel, argued that the Cry9C protein produced in *E. coli* might not be "properly folded." (Proteins are three-dimensional structures, and "misfolding" can lead to a loss of normal function within an organism.) In addition, he contended that Aventis officials had not established beyond a reasonable doubt that the Bt protein made by the *E. Coli* is the same as the Bt protein made by StarLink corn. People who did not react to the bacterial Cry9C might still react to the Cry9C found in StarLink corn.

Susan MacIntosh, an official with Aventis CropScience, countered that the Cry9C produced in *E. coli* is folded properly and functions the same as the protein produced in the StarLink corn. Both toxins are equally effective at killing European corn borers, she said. Aventis supplied the FDA with the bacteria-produced Cry9C that the agency had used in developing the blood test for humans. The Aventis practice of using bacteria-produced protein as a substitute for plant-produced protein for testing purposes was standard industry practice—and one long accepted by the EPA.

Keith Finger and Grace Booth, two private citizens who continued to believe they were allergic to StarLink corn despite FDA and CDC assurances to the contrary, spoke at the meeting. They asked the panel to seek further scientific inquiry into the FDA's blood test before allowing even trace amounts of Cry9C into the food chain.

Finger, an optometrist from Sebastian, Florida, told the panel that he had experienced allergic reactions to StarLink corn in commercial products on two different occasions. He had worked extensively in Central America and had eaten a great deal of corn-based food throughout his life without any problems. Finger said he first suffered an allergic reaction to what he believed was StarLink corn on September 14, 2000, after eating a meal of flour tortillas and beans and rice that contained whole corn solids. Four days later, on September 18, he read a news article about the Taco Bell taco shell recall and wondered if StarLink corn might have caused his allergic reaction.

Finger experienced a second allergic reaction in early April 2001, after eating three white corn chips that his wife had purchased at the grocery store. His wife had specifically bought white corn chips to avoid any yellow corn product that might contain StarLink corn. Following Finger's second allergic reaction, FDA officials tested a bag of white corn chips taken from the same store and found that they did indeed test positive for the DNA that codes for the Cry9C protein. The chips were the first known incident of StarLink contamination of a white corn food product in the United States.

His third reaction had taken place two days before the panel meeting, on July 15, 2001, under controlled conditions: he had performed an experiment to learn for certain whether he was allergic to StarLink corn. In the presence of medical personnel, he ate a small quantity of StarLink corn that had been sent to him, anonymously, in the mail after he had publicly stated his intention to eat StarLink in front of the EPA panel. Using a strip test, he had first confirmed that the corn kernels tested positive for Cry9C protein; then he ate about one-fifth of a tablespoon. The color slides he brought to the panel meeting showed the red welts that erupted on his skin after he ingested the StarLink corn. "My whole body was itching. You have no idea. The itching is absolutely intense," he told the panel. A physician diagnosed him as suffering from an allergic reaction and administered medication.

Finger stressed his belief that he had suffered allergic reactions to StarLink corn three times—and the third time he was certain he had eaten StarLink. "You all can sit here and say all you want about whether it causes allergy or not. But you're not going to change my mind," he told the panel. He asked the EPA not to approve any amount of StarLink corn in the food supply. "If I'm allergic to it, how many other people are allergic to it? We need a definitive test."

Grace Booth also spoke before the panel, describing her frightening experience in October 2000, when she had gone into anaphylactic shock after eating enchiladas made with yellow corn tortillas. She told the panel that she had sent a sample of her meal to FDA. "Nearly six months later, I was told by a form letter that there was no evidence that there was Cry9C protein in the tortilla," she said. "I'm told by the FDA and the CDC that there is no evidence that I am allergic to Cry9C, and I cannot claim with certainty that I am."

Booth visited an allergist who found she is not allergic to regular corn yet advised her to stay away from foods containing corn nonetheless. "So I wonder and agonize and live in fear of what I eat. I won't eat yellow corn products. And now I won't eat white corn," she said. Extending her arms, she told the panel that she would willingly donate blood to help scientists develop additional testing procedures. "To the panel and to the EPA, I plead, please, please do not allow me to be a casualty. Please take advantage of my willingness, my need to cooperate fully in further testing. Please do not approve StarLink until the questions are answered."

Dr. Hugh Sampson reiterated his belief that it was important to notify the medical community about the possibility that StarLink corn might cause allergies. Corn so rarely causes allergic reactions in people that it was "the absolute last thing" most allergists looked for, the pediatrician said.

The panel urged EPA officials to send e-mail messages about the potential allergenicity of StarLink corn to professional societies of allergists and other relevant lay groups. The scientists also said the federal agencies should screen farmers, millers, and other workers who have had extensive occupational exposures to StarLink corn for evidence of possible sensitivity to Cry9C.

Ultimately, the panel once again concluded that there was a "medium likelihood" that Cry9C protein is a human allergen. At the conclusion of the meeting on July 18, one scientist said, "I do not think things have changed since our last meeting."

Noting that the exposure of the U.S. population to StarLink has been "a massive, unintended natural experiment that deals with allergenicity and biotechnology," Dean Metcalfe urged the government to learn some lessons from the experiment. "We need to do an autopsy on the body of StarLink corn," he said. "EPA, FDA, and CDC, I strongly recommend, should form some kind of group to look at the problems—what went wrong, what went right."

On July 27, 2001, the EPA denied the Aventis petition.

FOURTEEN

FUTURE FOOD SECURITY

T he world received a reminder that genetic engineering is still in its infancy when the Human Genome Sequencing Consortium reported on February 11, 2001, that humans have only about 30,000 genes. Just eight months earlier, on June 26, 2000, the heads of the publicly funded Human Genome Project and the private Celera Genomics had announced with great fanfare at a White House ceremony that human beings have at least 50,000 genes and possibly as many as 80,000 to 100,000. (Later in the year, in August 2001, still other researchers disputed the 30,000 number as far too low and it may, in fact, be decades before scientists agree on the contours of the human genetic map.) But with scarcely twice as many genes as a worm or a fruit fly to our credit, many scientists began to posit that genes should be viewed more properly as a "guidepost" than as a "blueprint" for the makeup and functioning of human beings.

What does this portend for the genetic engineering of plants and animals? "It means that the old paradigm that one gene makes one protein is clearly in need of revision," an official with Rockville, Maryland-based Celera Genomics said.

This sudden turnabout in the thinking of mainstream genetic scientists was not lost on Mae-Wan Ho, who for years has insisted that the old paradigm of "one gene, one protein" was not supported by the scientific evidence. Ho and other critics of genetically engineered crops have long argued that genes within complex organisms interact with other genes and are influenced by their environment, making it impossible to predict with certainty how the insertion of foreign genes will affect a plant or animal.

"Genetic determinism is dead" and has been dead for nearly twenty years, Ho said in the February 2001 *ISIS News* online. "It is time to call a complete halt to all GM releases, and for scientists to go back to the drawing board, especially in accordance with the precautionary principle."

A few cutting-edge researchers have begun to do just that. "From a scientific perspective, the public argument about genetically modified organisms, I think, will soon be a thing of the past," Robert Goodman, professor of plant pathology at the University of Wisconsin in Madison, predicted at a meeting of the American Association for the Advancement of Science in San Francisco in February 2001. "The science has moved on, and we're now in the genomics era."

As scientists accumulate more knowledge about plant and animal genomes, a new revolution is in the offing, Goodman told me later. The future of improving crops lies in resorting and reshuffling the genes that are already in a plant's genome—not in plucking a gene from a foreign organism and trying to make it do something useful in the crop.

Goodman has been working in the field of genetic engineering for more than twenty years. He was vice president for research and development at Calgene when the company created the first gene-altered fruit, the Flavr Savr tomato. Instead of transplanting a gene from one organism to another in hopes of producing a desirable trait in a food plant, he predicts, scientists soon will be able to use genomics techniques to find out which clusters of genes within an organism are responsible for a particular desired characteristic. "You don't go out and get a single gene to fix something—you reorchestrate the expression of large numbers of genes in order to achieve a particular trait."

Goodman characterized the central dogma of genetic engineering— that one gene equals one protein equals one trait—as "an approximation that is vastly oversimplified." Genomics, he said, is paradigm shifting. "It's a totally different way of thinking. Instead of thinking gene by gene, we're thinking what I sometimes call 'massively parallel gene expression analysis.'" Scientists now talk about gene products in-

fluencing other genes, which in turn—given a particular environment
—cause other genes to express products. It is this complex cascading
interaction of genes, gene products, and environment that ultimately
results in observable traits, he said.

I asked Goodman if he believes recent discoveries about the nature
and role of genes mean that the technology of gene splicing is inherently limited.

"Yes, I think so," Goodman answered, "because we don't have
enough knowledge yet about the complex set of gene interactions that
are going to be required to achieve those traits. And that's something
people using genomics are going to figure out."

As evidence of the need for a paradigm shift, he pointed to the long
search for genes responsible for a trait in plants that scientists call *acquired disease resistance*. "Years ago people found a few proteins that
are always around when you have this heightened state of disease resistance," Goodman said, "and the genes for those were cloned. I did
some work on this ten, fifteen years ago and published papers, trying
to take individual genes and get this disease resistance phenomenon to
work. We were never successful.

"It turns out there are hundreds of genes that are involved, and we
can now see those genes by molecular tools," he continued. "We've
sequenced two hundred and eighty-six genes from rice that are associated with this disease-resistance response."

Genomics involves mapping entire arrays of gene expression that
are present, or active, in a plant or animal that exhibits a desired trait.
"It's called gene expression profiling. You're actually looking all at
once at ten thousand genes, or five thousand, or whatever the number
of genes that you've got on your array, and they're telling you what
the status of the expression of genes in that organism is." He predicts
that in the future, plant breeders will "basically be able to read the entire genome expression" and choose which members of a population
are going to be promising.

"The key is, you can detect function. You can see genes at work,
and you can focus on lots of genes all at once. This is what breeders
have done for more than a century," Goodman said. But they could
observe only the outward traits of the organisms that resulted from
crosses. "With new knowledge and modern tools of the trade, breed-

ers can make more rapid progress on many more traits than in the past."

"What do you think this new research will mean for genetic engineering as we now have it—in our soybeans and corn?" I asked.

"I have predicted this will all bypass genetic engineering within the next five to eight years," he answered.

Goodman's prediction that the first food crops made with gene guns will soon become obsolete is based partly on his sense that the public will not accept genetically modified foods, and partly on his belief that genomics is simply "better science" than gene splicing.

Donald Duvick, former director of plant breeding research at Pioneer Hi-Bred International, foresaw limitations to genetic engineering as a tool for improving food crops more than thirty years ago. "Recombinant DNA molecular research probably will have little direct impact on the development of useful new crop varieties," Duvick said in a talk delivered to the National Research Council in April 1977. "Complex, delicately balanced interactions among many genes determine the phenotypes of successful crop varieties.

"In most cases so-called 'single-gene' effects actually depend on a very complex background of supporting genes," he told the council. "Moving given genes from variety to variety often changes the intensity or even the nature of their effect; sometimes the gene produces unexpected deleterious effects." He further warned that when genes are moved from one species or genus to another, "unexpected and often deleterious effects can be even greater."

Wes Jackson, a geneticist and president of the Land Institute in Salina, Kansas, has also questioned the soundness of the science underlying genetic engineering. "Most of the promises that are being offered in biotechnology won't pay off. Ironically, though, in part because of what we're learning about the genome, money is being spent—and will be spent—chasing lots of wild geese, and there will be a smidgen left over that will be of use," he predicted during a debate on the risks and benefits of biotechnology sponsored by Environmental Media Services in Washington, D.C., in January 2000.

Jackson believes the emphasis now being placed on biotechnology in the United States stems from the failure of the government and academic institutions to invest money in the fields of ecology and evolutionary biology. "The people who are really promoting the biotechnology—the zealots—are fundamentally ignorant of the problem that the breeders have had to deal with over a long period of time," he said. Traditional breeding brings a lot of genes along that serve to protect the genomic structure of the plant or animal. "With some of this biotech, especially the transgene, where you're going to take the gene and park it in there, that gene mostly is going to fail."

In Europe too, a number of molecular biologists have had second thoughts about the direction of their research. "Some of the world's leading authorities who have pioneered genetic engineering are already saying that the revolutionary abilities of the science are inherently unstable," British MP Alan Simpson said in February 2000. "That is not to reject the science per se but to make the point that the possibilities of biotechnology are likely to be found only in the second generation of biotechnology—the ability to alter gene sequences within an organism rather than grafting sequences from one completely unrelated species to another. All of the evidence so far says that there is an inherent instability in that process."

Peggy Lemaux, a plant molecular biologist at the University of California at Berkeley and a proponent of biotechnology, has gently criticized the first generation of gene-altered foods forged by agrochemical companies. "The products in commercial production today represent only the first, rather crude attempts to use engineering to improve crop plants," she and a colleague wrote in a presentation to the California Weed Science Society meeting in Sacramento, on January 10, 2000.

"Glyphosate is considered low-risk for promoting herbicide-resistant weeds, but low-risk is not no-risk," Lemaux continued. So many farmers were planting Roundup Ready crops over such vast tracts of farmland that weeds would inevitably develop resistance to glyphosate, she warned. Widespread use of Roundup had already led to resistance in rigid ryegrass in the United States and Australia and in goosegrass in Malaysia.

"The first products of the technology are crude; the Roundup Ready soybean is not the best that can be done," she said. "New information gained from studies of the genome will provide new avenues for crop improvement that cannot be achieved in any other way."

In March 2001 scientists in Lemaux's laboratory unveiled what they describe as a better, safer way to genetically modify food plants. "Our method is ready for use by plant biologists to create new varieties of cereal crops in a manner that is both very quick and very safe," she said in a Berkeley press release. The technique will allow scientists "to boost nutritional content, improve pest resistance, reduce allergenicity and even perhaps speed up beer-making."

Peggy Lemaux and her colleague Thomas Koprek used mobile pieces of DNA known as "jumping genes" to deliver a desired gene into barley or other cereals: they engineered genes to "hitchhike" on the jumping genes. Grains were "surprisingly adept at stifling the implanted gene's activity," they found, "a process known as gene silencing." While their technique still employs genetic engineering, the scientists said the innovation "eases safety concerns and minimizes the problem of gene silencing."

To overcome gene silencing, the scientists in Lemaux's lab deliver just one copy of the gene of interest into the new plant genome, coaxing it into an area of the DNA that they consider favorable for making the desired protein. The technique is a refinement of earlier genetic engineering methods; scientists actually create two transgenic plants and then cross-breed them to get a transgenic plant that expresses the desired trait. In the first transgenic plant, they hitch a single copy of the gene of interest to the part of the jumping gene that allows it to move about in the genome. To the second plant, they add an enzyme that helps genes jump. In subsequent generations of the plant, after the introduced trait becomes integrated, the enzyme gene naturally breeds out of the plant, thereby shutting down the ability of the inserted gene to continue to jump about.

While some plant molecular biologists may begin to employ jumping genes or jettison gene splicing altogether in favor of genomics, many scientists and biotechnology companies plan to keep mining the older techniques. Monsanto certainly sees a role for genomics in its plant breeding program, but it has no plans to abandon the products and bioengineering methods it pioneered and patented in the 1980s and 1990s. "We are successfully integrating our experience in chemicals, seeds, genomics, and biotechnology," Hendrik Verfaillie, Monsanto's new president and CEO, wrote in a March 1, 2001, letter to share owners. Verfaillie, who worked his way up through the management structure, was product manager for Monsanto's weed-killers Lasso, Ramrod, and Machete, and served as vice president of the Roundup division.

Monsanto's future focus, Verfaillie said, will be "entirely on agriculture," which "offers a compelling opportunity for growth," he explained. He plans to expand the company's profitable line of Roundup Ready crops, which helped boost sales of the weed-killer to $2.6 billion in 2000, accounting for 47 percent of Monsanto's total revenues of $5.5 billion.

Roundup Ready soybeans are enormously popular with American growers because they simplify weed control, freeing up more time for other pursuits. They also offer producers a sizable boost in economic returns over conventional soybeans. Agricultural economists have calculated that even after taking into account the added cost of Monsanto's technology fee, growers planting Roundup Ready soybeans enjoy substantial savings per acre in reduced herbicide costs, fertilizer inputs, and labor costs over those using traditional soybeans. The USDA's annual grower survey, conducted in late March 2001, found farmers planting 76 million acres of soybeans—the largest area of soybeans ever cultivated in America. For the 2001 growing season, farmers planned to grow conventional, nonengineered soybeans on only 28 million acres and Roundup Ready soybeans on 48 million acres, accounting for a record 63 percent of the U.S. crop.

As the scientists who reengineered our crops reconsider the theoretical underpinnings of their methods, what are consumers to think about the "crude" genetically engineered foods lining our kitchen cupboards? Public health officials the world over have started to take a closer look at the presumed safety of the first generation of gene-altered foods for both humans and the ecosystem. In the judgment of many, the FDA and other regulatory agencies failed the public by approving gene-altered corn, potatoes, soybeans, and canola before thorough studies of the foods' long-term health effects were conducted. Ultimately, genetically modified crops may turn out to be completely safe for people to eat and cause no damage to wildlife, but critics contend that the federal government has not taken stringent measures to ensure their safety.

Canadian regulators too have come under attack for their hasty embrace of the first fruits to issue from biotech industry labs. Three countries—the United States, Canada, and Argentina—grow more than 95 percent of the world's genetically modified crops. In 2000 about 55 percent of the Canadian canola crop was genetically engineered for herbicide tolerance. But in February 2001 the Royal Society of Canada released a report that raised serious questions about the government's approval of gene-altered foods. Following a year-long review, a panel of fifteen independent scientists chastised Health Canada and the Canadian Food Inspection Agency for allowing genetically engineered crops to be channeled into human food in near secrecy.

Canadian regulators, the panel said, need to take a more precautionary approach to biotech food reviews. The burden of proof must be on the biotechnology companies to "carry out the full range of tests necessary to demonstrate reliably that they do not pose unacceptable risks." Like the British Medical Association, the Canadian scientists roundly rejected the U.S. FDA's concept of "substantial equivalence" to exempt gene-altered foods "from rigorous safety assessments on the basis of superficial similarities."

Brian Ellis, a biochemist and biotechnologist at the University of British Columbia and co-chairman of the Royal Society review panel, credits Canadian regulators with doing the best they could with the knowledge that was available ten years ago. But he also believes "there were clearly gaps" in the evaluation process.

"I would have to say that the chances of there being a serious negative impact are very, very small. Nevertheless, the fact that the very, very small chance never materialized, so far, you would have to just put down to luck," Ellis told me shortly after the report came out.

"There were uncertainties about the possible safety impacts either on humans or on the environment that would only get resolved over time. You don't have all the answers. You don't know how it will turn out until you've done the long-term tests," he added. "From a scientific point of view, that's a perfectly acceptable scenario. Now, if the public good, or the public safety is your ultimate goal, then you have a different standard of acceptability of risk."

New technologies always coexist with uncertainty, Ellis continued, drawing parallels between the current experiment with genetically engineered foods and the early days of nuclear power development. With nuclear plants, "there were issues that would only be resolved over time, with experience. The same thing will happen, I suspect, in the GM food industries as well. They'll sail through relatively unscathed for some period of time—maybe quite a bit of time—and sooner or later something will pop up that they didn't predict, that they hadn't imagined would be the case. Then they'll have to deal with that retrospectively."

The fifteen scientists from the Royal Society panel called on the government to shore up safety tests of gene-altered foods and open the approval process to public scrutiny. They also urged Canadian officials to establish clear rules about what types of toxicology studies companies should have to conduct to demonstrate that completely new constituents in gene-altered food plants are safe for people to eat. The panel advised that antibiotic resistance markers should not be used in gene-altered plants intended for human consumption. They also urged Health Canada to carry out "after-market surveillance of GM foods incorporating any novel protein."

The reviewers decried the fact that independent scientists who might critically review the safety of the new foods are vanishing. "The increasing domination of university research by the commercial interests of the researchers and their industry partners removes incentives for reliable scientific research on the safety of these products."

Like Robert Goodman, Brian Ellis believes the future of crop improvement lies in genomics. "One of our recommendations is that Canada develop a world-class genomics resource for each of its major crops and farm animals because that is the way the technology is going," he said.

"Do you think genomics will actually supplant gene-splicing technologies?" I asked.

"I think genomics is exactly where it will go. It's much more so-phisticated," he responded. "It's reorganizing. It's doing what classical breeding is doing, but giving you tools to monitor those breeding-induced changes very, very closely so that you can pick out the prog-eny from every cross that has the ideal combination of alleles. Then you're dealing with a stable gene structure, a genome structure that's directly analogous to what you get in classical breeding."

There are two ways of looking at the instability associated with ge-netically engineered plants, according to Brian Ellis. "One—the way industry looks at it—says, 'Well, I put that gene in there, and it's been through ten generations, and it's showing up just fine.' That's one kind of argument, and there's no doubt those genes are stable. But when I say 'instability,' it's a little fuzzier. What I'm talking about is that those new individual genes can have a lot of collateral effects within the genome, so they will change gene expression patterns in other genes," he explained. "It's change that you don't anticipate and can't antici-pate, so you have to find out by empirical observation and testing." The unforeseen response of Monsanto's Roundup Ready soybeans to hot weather conditions in Georgia, when farmers found their plants' stems splitting open, is an example of that kind of instability.

"I don't think you can think of all these things," Ellis said of the ge-netically engineered food crops on the market. "It's a five-year experi-ment—not just in terms of human health but in terms of how will they stand up in all possible environmental conditions, in all possible growing areas. Nobody can afford to do it as a pure trial beforehand, because that's an enormous undertaking. So how do you do it? You put them out in the fields. Now, in terms of how well the crop stands up, you can say that's the company's problem. If it crashes finally in Georgia on a forty-degree-centigrade day, the company has a prob-lem. But it's nobody's problem but theirs and the growers'.

"But when it comes to issues around food safety," he continued, "or impacts on human health, or collateral effects within the environ-ment, then it becomes everyone's problem."

While some scientists and health organizations have been reviewing the safety of gene-altered crops for humans, others have focused their investigations on the impact of gene-altered crops on wildlife. Scien-tists studying the effect of Bt corn pollen on monarch butterflies con-cluded in September 2001 that monarch caterpillars in the wild face

negligible risk from the most widely planted varieties. Extensive ecological studies published in the *Proceedings of the National Academy
of Sciences* in September 2001 concluded that the two Bt corn varieties most popular with American farmers, MON810 and Bt11, do
not contain high enough levels of the bioengineered toxin to endanger
monarchs. The studies were the result of a collaborative research
effort by industry, university, and USDA scientists (many of whom
presented preliminary findings to the monarch butterfly research symposium held in Chicago in November 1999).

The researchers did find that the pollen from one variety of Bt corn,
Event 176 marketed under the name KnockOut, is harmful to the
black swallowtail butterfly. Pollen from Event 176 corn "causes mortality and sublethal effects, such as growth inhibition, at very low
concentrations," Mark Sears and his colleagues wrote in a paper summarizing the comprehensive two-year Bt corn and butterfly review
effort.

Jane Rissler of the Union of Concerned Scientists said the studies
showed that the United States "has a regulatory system based on luck."
EPA had not discovered the danger to butterflies in the variety when
the agency approved Event 176. Fortunately, because the KnockOut
variety, made by Syngenta (formerly Novartis), accounted for only
about 2 percent of the annual corn crop, however, the effect on butterflies was deemed by scientists to be insignificant. Syngenta told the
EPA that it will phase out the variety by the end of 2003.

In October 2001, following what agency officials said was a comprehensive scientific review of the human health and environmental
safety concerns raised by many consumers and environmental groups,
EPA renewed the registrations of Bt corn for an additional seven
years. "Bt corn has been evaluated thoroughly by EPA, and we are
confident that it does not pose risks to human health or to the environment," EPA assistant administrator Stephen Johnson said in an
October 16 press release. "Consumers should be assured that these
corn varieties show no signs of any adverse effects to human health."

The Consumer Federation of America commissioned a review of the
safety of gene-altered foods in 2000. Their hefty report, released in
January 2001, confirmed that the FDA's oversight of genetically mod-

ified foods contained unsettling loopholes. There is "effectively no federal standard for human safety" applied to genetically engineered foods, the federation found. The group charged that this "laissez faire approach" to regulating gene-altered food frees the FDA from blame if the foods are later found to be unsafe.

A blue-ribbon panel assembled by the United States and the European Union to review the risks and benefits of agricultural biotechnology also urged the U.S. government to require stricter reviews of gene-altered foods. In a report released on December 18, 2000, the diverse group of twenty scientists, consumer activists, and farmers from both sides of the Atlantic called for mandatory labeling. "Consumers should have the right of informed choice regarding the selection of what they want to consume," the joint U.S.–EU panel said. Since the EU already requires such labeling, the advice was directed specifically at the U.S. government.

The group, which included Norman Borlaug, winner of the Nobel Peace Prize in 1970 for his work on the Green Revolution and a passionate advocate of biotechnology, called for "mandatory monitoring" and the ability to trace the presence of genetically modified products—neither of which is now possible in the United States. In assessing what was known about the safety of biotech foods on the market, the panel concluded: "At the present time, no obvious health effects have yet been identified with crops or foods that have been approved. Anticipated effects are likely to be of low-level, evident only after long periods of use among especially at-risk population groups, difficult to detect with certainty, and thus monitoring for such effects is likely to be costly to implement."

The contamination of much of the U.S. corn supply with the Cry9C protein in StarLink exposed the lack of a mechanism for recalling tainted gene-altered food products. The U.S.–EU panel emphasized that a capacity to trace these food products is "essential to ensuring consumer choice, understanding the causes and establishing liability in cases of unanticipated negative effects, ensuring effective product recall should a safety problem arise, and, in some cases, validating benefit claims."

The American Medical Association issued a report in December 2000 endorsing continued use of the FDA's concept of "substantial equivalence" in regulating genetically engineered foods. Federal regulatory oversight should continue to be "guided by the characteristics of the plant, its intended use, and the environment into which it is to be introduced, not by the method used to produce it, in order to facilitate comprehensive, efficient regulatory review of new genetically modified crops and foods," the AMA Council on Scientific Affairs stated.

While generally supportive of the U.S. regulatory scheme, the AMA suggested that some improvements were in order. "Methods to assure the safety of foods derived from genetically modified crops should continue to be refined and improved," the report concluded. "Although no untoward effects have been detected, the use of antibiotic markers that encode resistance to clinically important antibiotics should be avoided if possible."

The FDA continues to cite the apparent lack of adverse health impacts among 280 million Americans as evidence that gene-altered foods are safe. So far there are no obvious reasons to suppose that the population as a whole is either better or worse for their dietary exposure to gene-altered corn and soybeans. But the argument is not robust enough to persuade other countries to take a chance on these foods.

Some physicians worry that increased reliance on Roundup Ready crops—and increased use of the weed-killer glyphosate—might contribute to other disease problems. Two Swedish oncologists have drawn a connection between exposure to pesticides such as glyphosate and an alarming rise in non-Hodgkin's lymphoma. Hodgkin's disease and non-Hodgkin's lymphoma are two forms of cancer that afflict the lymphatic system. Since the early 1970s, the American Cancer Society has reported a dramatic 80 percent increase in the incidence of non-Hodgkin's lymphoma. Lennart Hardell and Mikael Eriksson published a study in the March 15, 1999, *Journal of the American Cancer Society* in which they suggested that exposure to glyphosate and other herbicides increases the risks for non-Hodgkin's lymphoma. The rapid increase in the use of glyphosate deserves further epidemiological study for such a possible link, the oncologists wrote. Other scientists, however, dismiss this suggestion, countering that evidence to support such a link is inadequate.

In Ireland some doctors have noticed an increase in soy allergies among children. Elizabeth Cullen, a public health physician and co-chair of the Irish Doctors' Environmental Association, says that at this point the observation is purely anecdotal. "Unfortunately, we've no systematic way to trace the cause," she told me. "It could be due to something else entirely, unrelated to soya, but the suspicion is there, in our opinion, that it could be due to genetically engineered foods." She added, "Anything with soya from America I would have to view with suspicion."

The Irish medical group is seeking a moratorium on genetically modified food constituents until they are properly tested. "If this were a drug, we feel it wouldn't be licensed," Dr. Cullen added. "In the absence of a moratorium, we want a register of diseases thought to be due to GM foods established.

"People tell me there's no evidence that genetically engineered foods cause problems," she told me. "Of course, there's no evidence, because nobody's collecting it. That's our big concern. We are concerned about possible adverse health effects and a complete absence of ways of checking for possible health effects." On April 19, 2001, the Irish Medical Organization passed a motion requesting that the minister for health and children provide funding to create "a group to establish the implications of genetically modified foods."

Many public health officials and scientists around the world believe that the long-term safety of gene-altered foods will have to be judged one food at a time. Even if Bt corn and Roundup Ready soybeans turn out to be completely safe, they argue, new varieties of gene-altered foods might present unexpected health problems.

Tewolde Berhan Gebre Egziabher of Ethiopia's environmental protection authority has argued that people should not eat genetically modified foods until scientists are absolutely certain they are safe—in another ten years, perhaps. "One genetically engineered product that is consumed and hasn't shown any demonstrable problems so far doesn't mean the next one will be the same," he told *New Scientist* in an interview January 20, 2001. Ethiopian officials are watching America's biotech food experiment to see how it turns out, before deciding whether the technology offers long-term benefits—or problems.

Meanwhile, on May 10, 2001, China's State Council passed a draft regulation requiring clear labeling of food products made with genetic alterations, China Online reported. The purpose of the legislation was to provide proper labels so that countries could decide for themselves whether to import genetically engineered foods. A news report out of Sri Lanka on May 13, 2001, said the government had decided to retain a ban on genetically modified foods that took effect May 1—despite sharp criticism from the United States.

In contrast with these actions abroad, legislation to label gene-altered food in the United States foundered. The Genetically Engineered Food Right to Know Act of 1999, introduced by Representative Dennis Kucinich, picked up support from fifty-six members of Congress. The legislation was offered as an amendment to the fiscal year 2001 agriculture funding bill but was withdrawn without a vote. Kucinich noted that "support in the House of Representatives does not parallel the support for labeling by the American public." He planned to reintroduce the bill in the fall of 2001 but delayed doing so in the immediate aftermath of the devastating September 11 attacks on New York and Washington, when the Congress turned its full attention to waging a war on terrorism.

On January 18, 2001, two days before the inauguration of President George W. Bush—and eight years after its infamous 1992 biotech policy was issued in the waning days of the senior Bush presidency—the FDA proposed formal regulations for bioengineered foods. The rule would require biotech food makers to give the FDA 120 days' notice before marketing a new genetically engineered food. It also would make mandatory the longstanding voluntary "consultations" between developers of gene-altered foods and the FDA. These mild proposed changes to the current policy were easy for the food industry to swallow; they have received the backing of the Grocery Manufacturers of America.

The FDA's proposed rule is most remarkable for what it omits: it requires neither mandatory labeling nor pre-market safety testing of genetically engineered foods. Rejecting the advice of independent scientists, rebuffing tens of thousands of Americans who testified at public hearings and wrote letters to the government, and disregarding sentiments voiced during the agency's own focus groups in 2000, the FDA again bowed to the interests of the biotechnology companies and food manufacturers.

Instead of mandating labels, which would bring America into line with the European Union, Japan, Australia, Thailand, Korea, and China, the FDA proposed voluntary labeling guidelines. The wording on the voluntary labels would simply allow a company to indicate "whether foods have or have not been developed using bioengineering."

The FDA acknowledged that the overwhelming majority of the 50,000 Americans who wrote to the government following the three public meetings held in the fall of 1999 asked for "mandatory disclosure of the fact that the food or its ingredients was bioengineered or was produced from bioengineered food." However, the agency justified ignoring those comments on the grounds that Americans had failed to provide a list of specific problems attributable to long-term ingestion of the foods. The government apparently expects clairvoyance of its citizens, since by definition "long-term consequences" may not become manifest for fifteen to twenty years or more.

Americans were basing their desire for caution on the outcomes of earlier, documented tragedies that followed the introduction of new technologies. Four years into the DDT experiment, for example, no one could have told the FDA that the pesticide would bring the bald eagle to the brink of extinction twenty years later.

The FDA's voluntary labeling scheme also prohibits organic growers and other food processors from calling their foods "GMO free" or "GM free" on a label. "There is potential for the term 'free' in a claim for absence of bioengineering to be inaccurate," the FDA explained. "Consumers assume that 'free' of bioengineered material means that 'zero' bioengineered material is present."

The FDA went on to suggest, in tortured language, that genes from biotech seeds and pollen have invaded American farmlands to such a great extent that no food producer can claim its foods are free of these ingredients: "Because of the potential for adventitious presence of bioengineered material, it may be necessary to conclude that the accuracy of the term 'free' can only be ensured when there is a definition or threshold above which the term could not be used."

The use of the term "adventitious"—the chance or accidental presence of a gene patented by Monsanto, Aventis, or Novartis—acknowledged, for the first time, that the U.S. government has allowed traditional crops and organic foods to become hopelessly contaminated by pollen and seeds from genetically engineered crops.

By late October 2001, the FDA had received 94,000 public comments on the proposed rule. Linda Kahl, a consumer safety officer in the FDA's Center for Food Safety and Applied Nutrition, said that it took the agency seven months just to log in the massive number of comments. She added that it was impossible to predict when a final rule will be written.

The USDA's Advisory Committee on Agricultural Biotechnology devoted two days in mid-April 2001 to a discussion of this new and pressing problem across America: ubiquitous contamination of conventional crops with "adventitious transgenic material"—gene-altered organisms. The heartland has become a genetic jumble. Genetically engineered crops have popped up as "volunteers" in fields where they were not intentionally planted. Bags of biotech seeds have been spilled off trucks or been mixed with conventional seeds. Genetically engineered seeds have found their way into bins of conventional crops and fallen into the millions of cracks and crevices in the combines, conveyor belts, and transport vehicles that move grain.

Organic farmers, conventional growers, and food manufacturers told USDA officials at the meeting in Washington, D.C., that it has become nearly impossible in America to grow, buy, or sell corn and soybeans that are not contaminated with biotech genes. Edward Korwek, an attorney with the Washington, D.C., law firm of Hogan & Hartson, served on the advisory committee, which was established by Agriculture Secretary Glickman in the spring of 1999. Korwek said that millers and food processors were still trying to negotiate contracts with growers for pure, nonengineered corn and soybeans, something impossible to guarantee. "On the topic of zero, I think it's important to realize that biology does not recognize unapproved or approved" varieties of genetically engineered crops, he said.

Michael Yost of the American Soybean Association and the American Oilseed Coalition told the USDA that farmers were being asked to sign contracts stipulating that nonengineered soybeans would be completely free of gene-altered organisms—something growers could not guarantee. He said the USDA should try to get the word out to producers and consumers that it is "unreasonable" to expect zero contamination.

Contamination of U.S. seed corn by illegal StarLink corn has become widespread—despite the expenditure of more than $90 million by Aventis, and millions of dollars more by the U.S. government, to locate the corn and take it out of circulation. In March 2001 the USDA had set up a buyback program for corn seed tainted with Cry9C protein. It had contacted 282 small seed companies and asked them to test their seed corn for Cry9C. By mid-April 2001, seventy-eight seed companies had told the USDA that some of their seed corn had tested positive for Cry9C, according to acting Deputy Undersecretary Hunt Shipman. The USDA expected to pay small seed companies between $20 and $35 million to destroy the contaminated seeds.

Few analysts believed that the USDA buyback program would completely eradicate the rogue gene from successive generations of seed corn. The USDA's own tests of thousands of grain samples turned up traces of Cry9C in 10 to 22 percent of the nation's corn, according to various department sources.

In the fall of 2000, as a glut of StarLink-contaminated corn inundated the domestic animal feed market, Iowa corn growers lost twenty cents a bushel on the price of their grain. Millions of bushels of excess corn are now in storage. Analysts say growers across most of the corn belt have the largest carry-over of corn in eight years. Grain exports to Asia have slid, and U.S. agricultural exports have declined since 1996. According to the USDA, total agricultural exports in 2000 were valued at $51 billion, down from about $60 billion in 1996.

Consumers in America and abroad are turning in ever greater numbers to organic products in a desperate attempt to avoid eating genetically engineered foods. But genetic pollution, or genetic "trespass," as some growers call it, is especially hard for organic farmers to avoid. Their small plots are often surrounded by huge tracts of land shedding gene-altered pollen.

Mary-Howell Martens and her husband Klaas farm eleven hundred acres of certified organic crops in Penn Yan, New York. Principally, they grow soybeans, corn, small grains, vegetables, and red kidney beans. "In our case, the market does think, still, that zero tolerance is a reasonable demand, which from my perspective is going to hurt the organic farmer more than anyone else, because it isn't reasonable, probably, under most conditions," she said.

Testing companies in Iowa have performed DNA studies on hundreds of samples of nonengineered corn grown by diligent organic farmers throughout the Midwest and found traces of contamination in all of them, Martens told the USDA. Seed supplies are arriving on organic farms already contaminated with traces of DNA from genetically modified varieties. She has urged organic farmers to try to produce as much seed as they can on-farm, or to buy seed from other organic farmers. Still, as many organic growers lament, there are no walls high enough to keep out the birds, bees, and wind carrying pollen and seeds from neighboring gene-altered crops.

The Wall Street Journal tested the extent of contamination of organic foods as well. As part of a major investigation reported on April 5, 2001, the newspaper purchased twenty health food products claiming to be free of genetically modified ingredients and sent them to independent laboratories for testing. Eleven of the twenty foods tested positive for trace levels of genetically engineered materials. Five of the food products contained "more substantial amounts."

The test results obtained by *The Wall Street Journal* confirmed the worst fears of consumers who are trying to eat a diet free of the new biotech foods. "Genetically engineered crops are a direct assault on the organic division of the natural products industry," Craig Winters of the Campaign to Label Genetically Engineered Foods charged. Labels saying a food is "GMO free" are probably misleading "because it is now virtually impossible to obtain organic foods that are a hundred percent free of pollution from genetically engineered crops," he acknowledged regretfully. "But we also feel it is a sad state of affairs that our government has allowed this genetic pollution to occur in the first place because of their faulty regulations."

Wheat is Monsanto's next target. The company has announced its intention to roll out Roundup Ready wheat sometime between 2003 and 2005. European consumers have adamantly stated they do not want genetically engineered wheat for their bread. The Japanese have clearly said they do not want it for their noodles. On February 22, 2001, Reuters reported that board members of the Japan Flour Millers Association adopted a position on genetically engineered wheat. "Japanese consumers are highly suspicious and skeptical about safety of GM farm products, which may be hazardous to human health and environment," the statement said, adding that "flour millers strongly doubt that any bakery, noodle and confectionary products made of GM wheat or even conventional wheat that may contain GM wheat will be accepted in the Japanese market."

Most Canadian wheat farmers have said they do not want Monsanto to introduce Roundup Ready wheat plants into Canada's fields. Wheat is a grass, like corn. It typically self-pollinates within its own head, but with wind and air movement, growers say, some level of cross-pollination can be expected. Genetic contamination of conventional wheat would inevitably occur, and most Canadian farmers do not want to risk losing their crucial export markets in Europe and Asia.

Many U.S. wheat producers too are wary of genetically modified wheat, fearing it will destroy export markets. Half of the wheat grown in the United States is sold abroad.

Wheat is a $4 billion industry in North Dakota. In February 2001 Assemblyman Philip Mueller introduced legislation into the state legislature calling for a moratorium on the sale, distribution, or planting of gene-altered wheat seed in the state until such time as Canada registers the grain for commercial sale; the bill contained a sunset date of July 31, 2003. Monsanto mounted an aggressive campaign against the proposed legislation. The assembly amended the bill to a study resolution calling for a full analysis of the economic impact of genetically modified wheat on North Dakota growers. When the resolution moved to the state senate, lawmakers in that body stripped out specific language about genetically modified wheat seed. The final, watered-down resolution called for a general study on the effects of all kinds of genetically engineered crops on the state's economy.

"I think the reason why the original legislation was changed was that a majority of the legislators agreed that they didn't need a resolution to restrict any kind of GMO product—that farmers and producers were well aware of their market, and weren't going to plant anything that would harm their market," Judge Barth, a marketing specialist with the North Dakota Wheat Commission, told me in May 2001.

"They were afraid that if they passed any kind of legislation that had moratorium language in it, that ongoing research wouldn't continue and we'd be at a disadvantage to other countries, to other exporters of wheat." He added that a Monsanto lobbyist had "threatened to pull research funding from North Dakota if the legislation went through."

Brian Ellis predicts that Roundup Ready wheat will be the crop that makes or breaks genetically engineered foods. Throughout the spring of 2001, gene-altered wheat was the hottest topic among participants at agricultural biotechnology meetings in Canada. "The regulators I talked to in Canada say 'We don't know why Monsanto's pushing this.' They've got everything to lose, because this could be the death knell of the whole technology—if the wheat market blows up," Ellis told me. "There was pretty good consensus that this was a major, major headache."

Judge Barth of the North Dakota Wheat Commission believes Monsanto will show restraint and respect the concerns of North America's wheat growers. "I really, truly think they have gotten the message," he said. "Of course, I can't say they're going to do the right thing, but I think they've received the message, and they're well aware of what's going on. I think they're concerned about doing the right thing."

I was not persuaded. "From the perspective of a company that makes $2.6 billion a year selling Roundup, perhaps pushing Roundup Ready crops is the 'right thing' to do. Isn't it naïve to believe the company will voluntarily hold back Roundup Ready wheat?" I asked him.

"There's no doubt. But I think what's holding it back is, there's not a huge rush from producers to want this technology," Barth responded. He stressed that the Wheat Commission is not opposed to the technology. Roundup Ready wheat just doesn't make sense from a practical, agronomic perspective. One economic study estimated that Roundup Ready wheat would result in savings of only about three dollars per acre over conventional varieties of wheat.

Growers in the northern part of the state often plant Roundup Ready canola one year and wheat the next. With this crop rotation, farmers can spray Roundup on the canola crop one year, which allows them to kill the toughest weeds in their fields. The following year they typically plant conventional wheat, to which they apply Puma or Discover. (Puma is a weed-killer produced by Aventis; Discover is made by Syngenta, a company formed by the merger of Novartis and Zeneca. Both herbicides kill foxtails and wild oats, which are grasses and have a similar growth pattern to wheat.) Growers in the southern part of the state often rotate conventional wheat with Roundup Ready corn or Roundup Ready soybeans, then follow a similar weed-management strategy.

In that sort of operation, producers don't see the need for Roundup Ready wheat. By using different herbicides every other year, Barth noted, farmers are able to break the weed cycle. "From the producer's side, that's why they're not saying, 'Hey, we want this.' We've already got some of these tools out there that we can use, and we have to be careful that we don't overuse this product and build up resistance."

John Ellis, a certified organic farmer in York, Nebraska, and his family raise vegetables, small fruits, and small grains on seventy acres at Libby Creek Farm. They mill red wheat and white wheat into whole wheat flour. "Our flour's good for baking bread," he told me. "It's all whole grain, not a lot of processing involved, so it's got the full nutrition in it."

Ellis worries that genetically engineered crops are destroying small farms by forcing farmers to sacrifice food quality in their quest for ever-lower production costs. He points to the case of corn, where foreign markets for U.S. corn have been disappearing one by one. The amount of corn available for domestic use has increased, driving down prices. "To say Monsanto can continue to develop genetically modified organisms is to say they can go ahead and make a sale on the seed and make lots of money on them, but it's the farmer that's losing his sales and his markets in the long run," he told me in April 2001. "Marketing and relationships to buyers and trust can take a long time to establish, and once that trust is lost, it may take a long time to get them back again."

In the spring of 2001, Ellis said he noticed more and more small farmers talking about food security. "It seemed odd to me to be hearing issues about food security that normally you wouldn't hear about, generally, unless we were in times of war or some real critical times," he said. "Yet just due to this type of technology, it's throwing us into discussions of food security. It is pretty important to keep the quality of food and to keep the nutrition in our food that we may be giving up for the bottom-line yield or ease of farmers' growing mass quantities. It's not to the benefit of health."

At this juncture, John Ellis sees gene-altering technology as doing more potential harm than good. "They need to get onto health, if that's the point of it. But if it's just to ease the growing conditions and help propagate industrial farming or food production, I'm not sure everyone wants to see their choices limited, that we go to all manufactured products that are laced with chemicals and really aren't for your health benefit," he said.

"The companies are thinking more about their bottom line. It seems more dollar driven. I could be wrong, I don't know their outlook on it, but that's what it appears," Ellis continued. "The food that is the end result—that really should be our top priority, because we're talking about our own health and the health of our communities and our families."

President George W. Bush proclaimed the week of May 13 through May 19, 2001, National Biotechnology Week. "Genetic engineering will enable farmers to modify crops so that they will grow on land that was previously considered infertile," Bush said, in remarks that could have been lifted from his father's old files—or from Bill Clinton's. "In addition, it will enable farmers to grow produce with enhanced nutritional value. Our nation stands as a global leader in research and development, in large part because of our successes in understanding and utilizing the biological processes of life."

C. Manly Molpus of the Grocery Manufacturers of America saluted the Senate on May 14, 2001, for passing a bipartisan resolution declaring the week National Biotechnology Week. "America's global leadership in biotechnology is producing healthier foods for consumers, earth-friendly farming methods, and a new source of hope for the world's hungry," Molpus said in a press release, drawing on the biotechnology industry's stock promotional phrases.

Behind the scenes, though, GMA—the stalwart and sometimes sole supporter of gene-altered foods—has been growing impatient with the biotech corporations. Big-name food companies want to see some tangible consumer benefits from the biotech venture.

"It's been all pain and no gain for the food industry," GMA vice president Gene Grabowski told me in May 2001. "We want to see an acceleration on the part of the biotech industry toward developing biotech ingredients and products that have direct consumer benefits that we can market on the label." Food makers have been holding out for the promised biotech goodies that will appeal to consumers. They are hoping to market a new product, for example, with a label that announces it contains "added lycopenes, enhanced through biotechnology," Grabowski said. "That's what we want to preserve, and that's the way we think labeling should work. Not the negative labeling that you get with a mandatory government label that says, basically, 'Warning, this product may contain biotech ingredients.'"

I asked him what enticing gene-altered foods were in the biotech industry pipeline.

"There are some experiments being done on apples and pears in Britain, right now, to include elements that would retard tooth decay. And then, of course, the golden rice and things of that nature. And natural decaffeination of coffee beans," Grabowski responded.

"Looking to the future," I asked him, "how do you think consumers in other countries, where they do have GM labels, will affect the viability of genetically engineered foods here?"

"Plenty of different things can happen, but the ultimate end will be that biotech will continue," he responded. "Because biotech is like the polio vaccine, it's like the jet engine, it's like the telephone. You're not going to stop it with legislation. It offers too many benefits to too many people to stop. You can't. It would be irresponsible to stop it."

Still, some food company CEOs are worried that their brands are at risk, he said. Campbell's soups and Kellogg's "Tony the Tiger" have been frequent targets of antibiotechnology activists. "It's our brand name foods that get boycotted and protested against," Grabowski complained.

"The root of it all," he added, "is that the food industry found itself in a situation where these ingredients were in our foods. We really didn't think about it. We didn't question it. The biotech industry was kind of quiet the way they introduced these ingredients in products, and so we found ourselves after a period of a few years with up to seventy percent of our products on the grocery shelf perhaps containing these ingredients," added Grabowski, who has been with the GMA for the past four years. "We were sort of stuck in the middle, if you will."

The U.S. food industry has grown edgy about its massive inventory of gene-altered foods. GMA companies have been checking the public's pulse with regularity. So far, they say, 67 percent of American consumers either haven't heard of genetically modified foods or don't know enough about them to form an opinion.

"When we started doing research on consumer reactions two and a half years ago, back in 1999, sixteen percent of people were adamantly opposed to biotech, and seventeen or eighteen percent were favorably disposed," Grabowski continued. "Flash-forward to today, with a margin of error of three point five percent, the needle hasn't moved—sixteen percent are still against it, seventeen percent are still for it. Now, the good news is, from our perspective, they have not been persuaded by the antibiotech protests, antics, and polemics to move against biotech. The bad news, from our perspective in the food industry, is, neither have they been persuaded to move forward and embrace biotech. So what demographers tell us is, people don't know enough about it, they're still learning about it, they don't feel they know enough to make up their minds."

Wild maize was first used and sown by Indians in Mexico between eight thousand and ten thousand years ago. American Indians began to cultivate corn some sixty-five hundred years ago. For Mexican and Pueblo Indians, corn is synonymous with life. It took a handful of global biotechnology companies less than ten years to put an indi-

gestible protein into corn kernels and make corn pollen toxic to butterflies. In November 2001, Ignacio Chapela and David Quist, researchers at the University of California, Berkeley, reported finding
genes from biotech corn in samples of native Mexican maize, or "cri-
ollo," gathered in a remote mountainous region of Oaxaca, Mexico.
The region is a center of origin and diversity for the world's corn. Researchers do not know how the native landraces became contaminated with biotech genes; the nearest region where bioengineered corn
was known to be planted is 60 miles from the Oaxaca fields. The
Mexican government imposed a moratorium in 1998 on the planting
of genetically engineered corn.

Wheat traces its lineage back more than ten thousand years, to the
mating of wild grasses in the Middle East. Stands of wild wheat grew
in abundance in the Fertile Crescent, where the cultivation of grain
and civilization blossomed hand in hand. The first farmers gathered
grain from stalks of wheat plants that had not yet shattered, scattering
their seeds on the ground. By the simple act of picking the seeds that
were the easiest to collect, then sowing some of those seeds the following year, the earliest farmers domesticated the plants. The resulting wheat plants lacked a natural way to disperse their seeds and came
to rely on humans for their survival.

Bread is sacred. Biologists and anthropologists have found that our
common ancestors selected and cultivated varieties of wheat most
suitable for making bread. Technology has changed farming practices
dramatically in the West since biblical times, but wheat has endured as
our staff of life. While most of the corn and soybeans grown in North
America today are fed to livestock, the entire wheat harvest is eaten by
people.

No one has asked for the chance to eat bread and pasta, bagels and
cakes, breakfast cereals and snacks made from wheat engineered to
tolerate glyphosate or other weed-killers. But it is not clear that government officials in the United States or Canada have the power or the
will to deny biotech companies a license to produce and sell gene-
altered wheat—unless consumers bring a lot more pressure to bear on
politicians than they have so far.

What if glyphosate-resistant wheat plants should inexplicably split
open under harsh conditions? What if gene-altered wheat should un-

expectedly harbor an unknown human allergen? What if some other totally unpredictable problem arises from gene-altered wheat ten or twenty years after it takes root on the Great Plains?

In ancient times people understood that seeds are the most critical link in the food chain. Our ancestors had a different horizon than the typical biotechnology company: they were looking not to the next quarter's profits, nor to the expiration of a biotech seed patent, but to the lifespan of their children, and their children's children. If industry pollsters are right, two-thirds of Americans are still in the dark about gene-altered food. Yet the decisions we make today, as a society, about the genetic alteration of crops are critical to our future food security.

NOTES

A WORD ON SOURCES

I reported on agricultural biotechnology and genetically engineered foods for four years, from May 1997 through 2001, during which time I attended more than a hundred federal government meetings, science symposia, public hearings, industry conferences, congressional hearings, and news conferences and briefings by environmental and consumer groups on this subject, all of which provided background, information, and insights for this work. I conducted interviews for *Food Chemical News* and *Pesticide & Toxic Chemical News* (both published by CRC Press, in Washington, D.C.), and when I quote them in this book, I generally cite the relevant article and publication in the endnotes. In researching this book, I interviewed scores of North American and European scientists, doctors, government officials, industry leaders, trade group officials, and members of environmental, consumer, and activist groups. Finally, in tracing the history and evolution of U.S. policies and regulatory oversight of biotech foods, I have relied primarily on government documents, some obtained under the Freedom of Information Act, symposia proceedings, court documents, and interviews with former government and industry scientists and officials.

PROLOGUE   THE FOOD EXPERIMENT

p. 3 *She used basic ingredients:* Grace Booth, remarks at a Genetically Engineered Food Alert news conference in Arlington, Va., July 17, 2001.

p. 4 *"a massive, unintended:* Remarks by Dean Metcalfe made during EPA Scientific Advisory Panel meeting in Arlington, Va., July 18, 2001.

p. 4 *"I believe that the:* Remarks by Marc Rothenberg made during EPA Scientific Advisory Panel meeting in Arlington, Va., July 18, 2001.

p. 4–5 *A bushel of soybeans yields about 48 pounds:* Monsanto submission to FDA, "Safety, Compositional, and Nutritional Aspects of Glyphosate-tolerant Soybeans: Conclusion Based on Studies and Information Evaluated According to FDA's Consultation Process," September 2, 1994, pp. 2–102.

p. 6 *About 80 percent of the annual corn harvest:* National Corn Growers Association statistics.

p. 6 *In February 1999:* International Food Information Council survey, "U.S. Consumer Attitudes Toward Food Biotechnology," February 5–8, 1999. Online at http://www.ificinfo.health.org.

p. 6 *"There's just not enough data:* This and following remarks from Martha Herbert, delivered to the National Academy of Sciences' Standing Committee on Biotechnology, Food and Fiber Production, and the Environment, May 5, 2000, and author interview.

p. 7 *In company-sponsored tests:* Monsanto provided information on animal feeding studies in "Safety, Compositional, and Nutritional Aspects of Glyphosate-tolerant Soybeans," pp. 46–53.

p. 8 *Fifteen food chains in the United Kingdom:* The letter, "Open Letter to U.S. Commodity Agribusiness from UK/European Retailers," was reproduced in full in *Pesticide & Toxic Chemical News,* June 4, 1997. In addition to the British Retail Consortium, the letter was signed by Danish Commerce & Services; Federation of Retail Grocers in Denmark; Federation of Finnish Commerce and Trade; Finnish Food Marketing Association; Fédération du Commerce et de la Distribution in France; BFS in Germany; Food and Agricultural Policy Council of Swedish Retail and Wholesale Trade; and the Swedish Food Retail Association. Companies signing the letter included Asda Stores, Ltd.; Carrefour Nederland BV; Coop Schweiz; Co-operative Union Limited; Deutsche SPAR Handelsgesellschaft; Groupe Comptoirs Modernes; ICA Handlarnas AB/ICA Forbundet; Kwik Save Group Plc; Metro AG; Prisunic Exploitation; Safeway Plc; J Sainsbury Plc; Somerfiel Stores Ltd.; Tesco Stores Ltd.; and Waitrose Ltd.

p. 8 *"They look alike, they taste alike:* Interview with Monsanto spokeswoman, "Europeans Renew Push for Labeling of Genetically Modified Soybeans, Corn," *Pesticide & Toxic Chemical News,* June 4, 1997.

ONE **FORCE-FED CONSUMERS**

p. 11 *In the early morning hours:* This account is based on a letter placed on the Internet by a self-described participant in the action, "An Open Letter from GELF: Digging the Beet Down in Carlow," online at http://www.enviroweb.org/shag/info/reports/action6a.html.

p. 11 *Patricia McKenna, Ireland's Green Party:* Kevin O'Sullivan, "McKenna Praises Genetic Beet Crop Sabotage," *Irish Times,* October 1, 1997.

p. 12 *Many law-abiding Irish consumers:* While support for the action is difficult to quantify, 3,500 citizens responded to a second application by Monsanto for a further ten sugar beet trials filed about a year after the first application. Most consumers voiced objections to the outdoor tests.

p. 12 *"The public is frightened by:* Clare Watson remarks, author interview.

p. 13 *"Ireland regards itself as:* Quentin Gargan remarks, author interview.

p. 13 *"Genetic Engineering: Too Good to Go Wrong?":* Douglas Parr, Greenpeace report, London, October 1997.

p. 15 *"There's a lot of unease:* Douglas Parr remarks, author interview.

p. 16 *In May 2001 the number of cases:* "Century for vCJD," *New Scientist,* June 2, 2001.

p. 17 *Food processing companies:* Jack Doyle, "Potential Food Safety Problems Related to New Uses of Biotechnology," *Proceedings: Symposium on Biotechnology and the Food Supply* (Washington, D.C.: National Academy Press, 1988).

p. 18 *the Chocolate Manufacturers:* Office of Technology Assessment, *New Developments in Biotechnology: U.S. Investment in Biotechnology,* July 1988, p. 206.

p. 18 *"Biotechnology is, without a doubt:* Theodore Hullar, *National Agricultural Biotechnology Report 5: A Public Conversation About Risk,* (Ithaca, NY: National Agricultural Biotechnology Council, 1993), pp. 13–15.

p. 18 *"The plant revolution is proceeding:* Philip Abelson and Pamela Hines, "The Plant Revolution," *Science,* July 16, 1999, p. 368.

p. 19 *"Because trade is so important:* Letter to President Bill Clinton from forty U.S. trade groups, June 18, 1997, posted online at: http://www.

pmac.net/tough.htm. Signatories to the letter included the American Farm
Bureau Federation, the Grocery Manufacturers of America, the National
Cattlemen's Beef Association, the National Association of Wheat Growers,
the National Food Processors Association, the American Seed Trade Association, the Biotechnology Industry Organization, and the biotechnology
companies AgrEvo USA, Monsanto, Mycogen, and Novartis.

p. 20 *Edward O. Wilson has estimated that three:* Edward O. Wilson,
The Diversity of Life (New York: W. W. Norton & Co., 1992), p. 280.

p. 21 *Stephen Baillie, an ornithologist with the British:* Stephen Baillie
and R. D. Gregory, "Farmland Bird Declines: Patterns, Processes and Pros-
pects," *British Crop Protection Council Symposium No. 69: Biodiversity
and Conservation in Agriculture* (1997), pp. 65–81.

p. 22 *Avery argued that the biggest threat:* Dennis Avery, "Saving the
Planet with Pesticides, Biotechnology and European Farm Reform," *British Crop Protection Council Symposium No. 69: Biodiversity and Conservation in Agriculture* (1997).

p. 24 *On the night of June 3, 1998:* Genetic Engineering Network,
press release, UK, June 4, 1998.

p. 25 *Genetic engineering "takes mankind:* Prince Charles, *Daily Telegraph,* June 8, 1998.

p. 25 *Under cover of darkness:* For an account of the crop destruction
at Kirby Bedon, see Nick Hopkins, "Sowing the Seeds of Dissent," *Guardian,* June 13, 1998.

p. 25 *On Saturday, July 4, 1998:* GenetiX snowball, press release, Man-
chester, UK, July 4, 1998. The genetiX snowball campaign of "civil responsibility" noted that during the peace campaign in the UK in the 1980s,
more than two thousand women were fined for publicly snipping the wire
at Greenham Common in protest of the government's placing cruise mis-
siles on the site.

p. 27 *Subsequent studies at the University:* "Research Shows Roundup
Ready Soybeans Yield Less," University of Nebraska News Service. The
undated news release said that based on a two-year study at the University
of Nebraska Institute of Agriculture and Natural Resources, "Roundup
Ready soybeans yield 6 percent less than their closest relatives and 11 percent less than high-yielding conventional soybeans."

p. 27 *In Germany environmental activists:* SHAG, press release, August 6, 1997, online at http://www.enviroweb.org/shag.

p. 28 *On January 27, 1997, hundreds of activists:* Reuters, January 28,
1997.

p. 28 *In Austria 1,226,551 citizens:* "Austria Says No to Franken-foods," Reuters, April 15, 1997.

p. 28 *José Bové, a passionate:* See http://home.intekom.com; for additional views of the Confédération Paysanne, see http://www.mygale.org.

p. 29 *"We believe customers need:* Kathleen Hart, "Industry Tackles Stiff Opposition to GM Foods in Europe," *Food Chemical News,* June 29, 1998.

p. 30 *"I'm used to eating certain foods:* Kathleen Hart, "European Consumers' Groups Seek Segregation of GM Crops," *Pesticide & Toxic Chemical News,* July 9, 1998.

p. 30 *"choices and information:* Ibid.

TWO   **ALTERED STAPLES**

p. 31 *On May 27, 1998, a coalition of rabbis:* Material on the lawsuit *Alliance for Bio-Integrity, et al. v. Donna Shalala, et al.,* filed in May 1998 in the U.S. District Court for the District of Columbia, is based on information presented by plaintiffs at their press conference announcing the lawsuit, in court documents, and in interviews with Steven Druker and Joseph Mendelson. See also Kathleen Hart, "Coalition Sues FDA to Require Labeling and Testing of Genetically Engineered Foods," *Food Chemical News,* June 1, 1998.

p. 34 *"There are thousands of genes:* See Mary Alice Sudduth, "Genetically Engineered Foods: Fears and Facts. An Interview with FDA's Jim Maryanski," *FDA Consumer,* January/February 1993.

p. 37 *"There is no scientifically valued:* Grocery Manufacturers of America, "Don't Turn Back the Clock on Enhanced Foods, GMA Says," news release, May 27, 1998.

p. 38 *"an agenda to make the critics:* Stephen Druker, author interview. Druker sent a letter to *The Wall Street Journal,* August 22, 1999, saying in part: "It is unprecedented for eminent scientists to have joined such a lawsuit, and it is surprising that your article failed to mention their presence and their concerns. Apparently, you decided to characterize our suit as religion versus science when in fact it is a case of respected scientists contesting a public policy on scientific grounds, in concert with religious leaders who contest the policy for spiritual reasons."

p. 38 *"the U.S. media blackout:* Ronnie Cummins, author interview.

p. 39 *"You don't have people controlling:* Joan Konner, "Of Clinton, the Constitution and the Press," *Columbia Journalism Review,* March/April 1999.

p. 39 *"Though some publishers and broadcasters:* Melody Petersen, "Farmers' Right to Sue Grows, Raising Debate on Food Safety," *New York Times,* June 1, 1999. The book to which Petersen refers is Marc Lappé and Britt Bailey, *Against the Grain: Biotechnology and the Corporate Takeover of Your Food* (Monroe, Me: Common Courage Press, 1998).

p. 39 *Winfrey said the information stopped her:* Janet McConnaughey, Associated Press, February 9, 2000.

p. 40 *"Biotechnology's been around:* Widely quoted remark by Dan Glickman, March 13, 1997. See "What the Experts Say About Food Biotechnology," http://ificinfo.health.org/foodbiotech/whatexpertssay.htm.

p. 41 *"There is a lot at stake in what is going on:* Letter from John J. Walsh, Cadwalader, Wickersham & Taft, New York, to Roger Ailes, chairman and chief executive officer, Fox News, New York, February 21, 1997.

p. 41 *The second letter:* Letter from John J. Walsh to Roger Ailes, February 28, 1997.

p. 42 *"Our dream turned into:* Steve Wilson and Jane Akre remarks, author interviews.

p. 42 *Some studies have suggested that cows:* The Scientific Committee of the European Union on Veterinary Measures later reviewed this subject in "Report on Public Health Aspects of the Use of Bovine Somatotrophin, 15–16 March 1999." The committee found, "Present data do not provide a conclusive answer to whether or not previously applied analytical techniques have underestimated the actual IGF-1 level in milk by neglecting the protein-bound fraction, and to what extent the ratio between free and bound IGF-1 milk has changed as a consequence of rBST treatment resulting in a relative increase of the free IGF-1 fraction. Application of rBST increases the amount of excreted IGF-1 in milk by 25–70% in individual animals. The Committee noted that bovine milk may contain truncated IGF-1 (des(1–3)IGF-1) which was found to be even more potent than IGF-1 in the anabolic response when given subcutaneously to rats." See the full text at http://europa.eu.int/comm.

p. 42 *"With the complicity of the FDA:* Samuel Epstein, "Unlabeled Milk from Cows Treated with Biosynthetic Growth Hormones: A Case of Regulatory Abdication," *International Journal of Health Sciences,* vol. 26, no. 1 (1996), pp. 173–85.

p. 43 Science *magazine reported that men with high:* June Chan et al., "Plasma Insulin-Like Growth Factor-I and Prostate Cancer Risk: A Prospective Study," *Science,* vol. 279 (January 23, 1998), pp. 563–66.

p. 43 *British journal* Lancet, *reported that premenopausal:* Susan Hankinson et al., "Circulatory Concentrations of Insulin-like Growth Factor I and Risk of Breast Cancer," *Lancet,* vol. 351, no. 9113 (May 9, 1998), pp. 1393–96.

p. 43 *"There is no difference:* Kathleen Hart, "rBGH Should Be Removed from Market, Groups Tell FDA," *Food Chemical News,* December 21, 1998.

p. 43 *The Canadian scientists raised many questions:* Health Protection Branch, Health Canada, "rBST (Nutrilac) 'Gaps Analysis' Report," April 21, 1998.

p. 44 *"an antibody response" to the drug:* Remarks by Michael Hansen given at a news conference in Washington, D.C., December 15, 1998.

p. 44 *Monsanto sent a letter to a staff member:* Letter dated December 14, 1998, from Monsanto to Jay Hawkins, Office of Senator Jeffords.

p. 45 *Phil Metlin "together with an army:* Lawrence K. Grossman, "Blowing the Whistle On Your Own Station," *Columbia Journalism Review,* March/April 2001.

THREE **PUSZTAI'S POTATOES**

p. 47 *"We are assured that this is absolutely:* This and all ensuing quotations from Arpad Pusztai are based on author interviews.

p. 49 *The Flavr Savr contained a copy:* Information on the Flavr Savr tomato is based on an author interview with Robert Goodman.

p. 52 *"The new health findings:* "Experiment Fuels Modified Food Concern," BBC News, August 10, 1998.

p. 53 *Clare Watson of Genetic:* Genetic Concern (Ireland), press release, August 12, 1998.

p. 53 *"It is now clear:* Kathleen Hart, "Researcher at Rowett Institute Ignites Public Concern in UK over Safety of Biotech Foods," *Pesticide & Toxic Chemical News,* August 20, 1998.

p. 54 *A Monsanto official gave the conference:* The remarks from the Monsanto official and from Martin Dickman and Ralph Hardy, as well as information on input and output traits, were given at International Business Communications' Third International Conference on Transgenic Plants, November 2–3, 1998, Lake Buena Vista, Florida. See Kathleen Hart, "Pest Management Future Is in Biotechnology, Not Chemistry, Experts Say," *Pesticide & Toxic Chemical News,* November 15, 1998.

p. 55 *Optimum, a high-oleic acid soybean oil:* Optimum Quality Grains promotional literature.

p. 57 *Deans and provosts from twenty-two major agricultural:* Signatories to the document, "Vision for Agricultural Research and Development in the Twenty-first Century: Biobased Products Will Provide Security and Sustainability in Food, Health, Energy, Environment and Economy," from the National Agricultural Biotechnology Council, Ithaca, New York, include deans, provosts, and other administrative officers from: Clemson University, Cornell University, Iowa State University, Michigan State University, North Carolina State University, Pennsylvania State University, Purdue University, Ohio State University, Oregon State University, Texas A&M University, University of Arizona, University of California at Davis, University of Florida, University of Georgia, University of Hawaii, University of Illinois at Champaign-Urbana, University of Minnesota, University of Missouri at Columbia, and University of Nebraska.

p. 58 *"We cloned a gene, that of an alfalfa:* Martin Dickman, author interview.

p. 59 *"It's really clear to me:* Philip Regal's remarks are contained in the transcript of the FDA public meeting in Oakland, California, December 13, 1999.

p. 60 *In general, researchers want to slow down:* Sally McCammon, author interview.

p. 60 *Plants, bacteria, and algae:* Ernest Callenbach, *Ecology: A Pocket Guide* (Berkeley: University of California Press, 1998), pp. 85–86.

p. 60 *"If we look back over the whole":* Ibid.

p. 60 *Some of the changes that researchers envision:* According to two lists provided to me by Sally L. McCammon, "Field Tests of Transgenic Plants with Product Quality Phenotypes (1993–1998)" and "Field Tests of Transgenic Plants with Agronomic Properties Phenotypes (1993–1998)," the USDA has granted permission for researchers to test fields of crops with altered: amino acid composition; carbohydrate metabolism; carotenoid content; ethylene metabolism; fatty acid levels; feed properties; fiber strength; fiber quality; fruit ripening; fruit sugar profiles; lignin biosynthesis; nitrogen metabolism; nutritional quality; oil profiles; oil quality; phytosterols; pigment metabolism; processing characteristics; protein; seed composition; seed quality; senescence; starch metabolism; and storage proteins; also plants genetically engineered with reduced cationic peroxidase; ACC oxidase; disulfides in endosperm; ethylene production; flavonoid levels; flower and fruit abscission; fruit invertase level; fruit pectin esterase

level; fruit polygalacturonase; lignin levels; polyphenol oxidase levels; seed
set; and steroidal glycoalkaloids. In addition scientists are experimenting
with designing plants that have increased: dry matter content; antioxidant
enzyme; fruit solids; fruit sweetness; fruit firmness; lysine level; methionine
level; phosphorus; protein lysine level; seed methionine storage; seed size;
starch level; tryptophan level; tuber solids; and vernolic acid in seeds. Scientists also are genetically engineering food crops for: reduced bruising; delayed softening; extended flower life; stunted growth; prolonged shelf life;
inhibition of mycotoxin production; enhanced photosynthesis; antho-
cyanin production in seed; heat stable glucanase production; and linolenic
acid production. The USDA also has approved field tests of genetically engineered plants with altered: maturing; development; hormone levels; carbon dioxide metabolism; calmodulin level; cell walls; growth rate; and
pigment metabolism. Genetically engineered plants have been tested with
increased: ammonium assimilation; carbohydrate level; chlorophyll; cutting rootability; growth rate; stalk strength; tuber number; stem length;
and tolerance to salt, cold, drought, aluminum, and stress. Still other experiments involve confidential business information and are not publicly
available.

p. 61 *In one of the earliest analyses:* Roger Wrubel and Sheldon
Krimskey, "Field Testing Transgenic Plants: An Analysis of the US Department of Agriculture's Environmental Assessments," *BioScience,* vol. 42,
no. 4 (1992).

FOUR   **WHO'S MINDING THE GARDEN?**

p. 64 *In 1973 Herbert Boyer:* While there are many accounts of this experiment, Eric Grace provides a clear explanation of this landmark discovery and other basic biotechnology concepts in *Biotechnology Unzipped:
Promises and Realities* (Washington, D.C.: Joseph Henry Press, 1997),
p. 42.

p. 64 *One of the first commercially successful:* For an account of the
first genetically engineered food processing enzymes, see Susan Harlander,
"Food Processing Biotechnology," *National Agricultural Biotechnology
Council Report* 2 (Ithaca, NY, 1990), pp. 145–48. The first genetically engineered food-grade microorganism, a strain of yeast used in the baking
industry, was actually approved for use in March 1990 by the British Ministry of Agriculture, Fisheries, and Food, according to Harlander. In March

1993 the U.S. FDA approved its first genetically engineered enzyme for use in food processing, called chymosin. Recombinant chymosin is widely used in the manufacture of cheese, to speed up the formation of curds. While traditional rennet is scraped from the forestomach of calves, recombinant rennet is produced by genetically engineered *E. coli* and then purified from the resulting fermentation broth.

p. 65 *"Knowledge, whether for its own sake:* Gerry Waneck, "Safety and Health Issues Revisited," *Science for the People,* vol. 17, no. 3 (May/June 1985), pp. 38–43.

p. 66 *In 1983 Stanley Abramson:* EPA memorandum from Stanley H. Abramson, associate general counsel, Pesticides and Toxic Substances Division, to John A. Todhunter, assistant administrator for pesticides and toxic substances, "Status of Recombinant DNA and New Life Forms Under TSCA," March 14, 1983. In the memorandum, Abramson presents the following justification for regulating recombinant DNA and new life-forms as chemical substances: "Any DNA molecule, however created, is an organic substance of a particular molecular identity and is a combination of organic substances of particular molecular identities occurring in whole or in part as a result of a chemical reaction or occurring in nature."

p. 66 *"This last year was:* EPA hearing, Waterside Mall, Washington, D.C., March 18, 1983.

p. 68 *"When people are confronted:* Al Gore, Jr., "A Congressional Perspective," in *Biotechnology Implications for Public Policy* (Washington, D.C.: Brookings Institution, 1985), p. 12.

p. 69 *"The issue of government:* Ralph Hardy and David Glass, "Our Investment: What Is at Stake?" *Issues in Science and Technology* (Spring 1985), p. 79.

p. 69 *Hardy, who became director of life sciences:* Ralph Hardy, author interview.

p. 69 *"Society cannot insist:* David Jackson, *Biotechnology Implications for Public Policy* (Washington, D.C.: Brookings Institution, 1985), p. 67.

p. 70 *In April 1983, following two RAC:* See *1992 National Biotechnology Policy Board Report* (Bethesda, Md.: National Institutes of Health, Office of the Director, 1992), p. E–10.

p. 72 *In September 1987 Gary Strobel:* Keith Schneider, "Tearful Scientist Halts Gene Test," *New York Times,* September 4, 1987, p. A1.

p. 75 *"Much of the mistrust of biotechnology:* R. James Cook, "Agricultural Biotechnology and the Public Good," *National Agricultural Biotechnology Council Report 6* (Ithaca, N.Y., 1994), p. 59.

p. 75 *Recombinant DNA technology, he wrote:* George Wald, "The Case Against Genetic Engineering," *Sciences* (September–October 1976); reprinted in *The Recombinant DNA Debate,* edited by David A. Jackson and Stephen P. Stich, Prentice-Hall, Englewood Cliffs, N.J., 1979, pp. 127–28.

p. 76 *"clear distinction between classical:* Kathleen Hart, "Coalition Sues FDA to Require Labeling and Testing of Genetically Engineered Foods," *Food Chemical News,* June 1, 1998. When scientists with an interest in accuracy write about biotechnology, they generally take care to differentiate between old uses of living organisms, such as selective breeding, and modern uses of organisms, such as recombinant DNA techniques. In a 1988 report from the Congressional Office of Technology Assessment, for example, the authors at the outset draw a distinction between the "new biotechnology," which they say includes "recombinant DNA techniques, cell fusion, and novel bioprocessing techniques," and the "old biotechnology," which encompasses "the use of micro-organisms for brewing and baking or selective breeding in agriculture and animal husbandry."

p. 76 *Through polls and surveys:* David Schmidt, "The Impact of Language on Consumer Acceptance of Food Biotechnology," presentation to the IBC USA symposium on genetically modified foods, Washington, D.C., November 10–12, 1999.

p. 77 *"No evidence based on: Introduction of Recombinant DNA-Engineered Organisms into the Environment: Key Issues* (Washington, D.C.: National Academy Press, 1987).

p. 77 *"Introduction of transgenes into plants: Genetically Modified Pest-Protected Plants: Science and Regulation,* National Academy of Sciences, 2000.

p. 78 *"Four Principles of Regulatory Review: 1992 National Biotechnology Policy Board Report* (Bethesda, Md.: National Institutes of Health, Office of the Director, 1992), p. E-6.

p. 79 *The policy stated that:* "Statement of Policy: Foods Derived from New Plant Varieties," *Federal Register,* May 29, 1992.

p. 79 *The journal Bio/Technology reported:* Jeffrey Fox, "FDA Dishes Up Food Policy," *Bio/Technology,* vol. 10 (July 1992).

p. 80 *"Our feeling was that the FDA policy:* Ralph Hardy, author interview.

p. 81 *"Companies are now ready:* Memorandum from FDA commissioner David A. Kessler to the secretary of health and human services, "FDA Proposed Statement of Policy Clarifying the Regulation of Food Derived from Genetically Modified Plants—Decision," March 20, 1992. This memorandum and the following comments from FDA scientists Linda Kahl and Louis Pribyl to James Maryanski were obtained from the court docket in the case of *Alliance for Bio-Integrity, et al. v. Donna Shalala, et al.,* filed in May 1998 in the U.S. District Court for the District of Columbia.

p. 82 *In 1990 the American Medical Association (AMA) joined:* Kathleen Hart, "AMA to Revisit Policy on Genetically Engineered Crops," *Pesticide & Toxic Chemical News,* September 2, 1999.

p. 84 *FDA scientist Edwin Mathews:* Mathews warned that "genetically modified plants could also contain unexpected high concentrations of plant toxicants. The presence of high levels of toxicants could be amplified through enhancement of toxicant gene transcription and translation. This might occur as a result of up-stream or down-stream promotion of gene activities in the modified plant DNA." He noted that "the task of analysis of all major toxins in genetically engineered plant food includes the assessment of both expected toxicants and unexpected toxicants that could occur in the modified plant food. The unexpected toxicants could be closely related chemicals produced by common metabolic pathways in the same plant genus/species; however, unexpected toxicants could also be uniquely different chemicals that are usually expressed in unrelated plants."

p. 84 *In potato plants:* Janet Andersen, March 24, 1999, testimony before a joint hearing of the Subcommittee on Risk Management, Research, and Specialty Crops and the Subcommittee on Department Operations, Oversight, Nutrition, and Forestry of the Committee on Agriculture, House of Representatives.

p. 85 *"During the consultation process:* "Guidance on Consultation Procedures Foods Derived from New Plant Varieties," FDA Center for Food Safety and Applied Nutrition, October 1997.

p. 85 *"Foods are not required to undergo:* James Maryanski, from FDA transcript "Public Meeting: Biotechnology in the Year 2000 and Beyond," Oakland, Calif., December 13, 1999.

p. 85 *"Monsanto should not have to:* Michael Pollan, "Playing God in the Garden," *New York Times Magazine,* October 25, 1998, p. 51.

p. 86 *"People can postulate lots:* This and ensuing remarks, author interview with James Maryanski.

FIVE   **GUNPOWDER AND CORN EMBRYOS**

p. 88 *On December 8, 1998, President Clinton:* White House Office of Science and Technology Policy, "President Clinton Announces Recipients of Nation's Highest Science and Technology Honors," news release, December 8, 1998.

p. 90 *"You could generate millions:* This and ensuing remarks by Ernest Jaworski are based on author interview.

p. 91 *In 1981 Monsanto entered into a collaborative:* This history of Monsanto's milestones and acquisitions is based on company literature and press releases, including timeline history of Monsanto at http:// www.monsanto.com/monsanto/about-us/company-timeline; and "Monsanto Completes Successful Season for Agricultural Biotechnology Products and Sets Stage for 1998 Growing Season," company news release, December 10, 1997.

p. 91 *He embarked on an aggressive mission:* Among the companies Monsanto acquired or entered into partnerships with were Agracetus, Asgrow Agronomics, Dekalb Genetics, Holden's Foundation Seeds, and Monsoy in Brazil.

p. 91 *In January 1998, when the company's life:* "Life Begins at 97," ad copy, *Washington Post,* January 22, 1998.

p. 91 *In 1975 two Flemish researchers, Jeff Schell:* Margie Patlak with Roger Beachy, Mary-Dell Chilton, Maarten Chrispeels, Nina Fedoroff, Robert Haselkorn, Ernest Jaworski, and Arthur Kelman, "Beyond Discovery: The Path from Research to Human Benefit" (Washington, D.C.: National Academy of Sciences, October 1998).

p. 91 *The device, also called a particle gun:* "Development of the 'Gene Gun' at Cornell," Communications Services news feature, New York State Agricultural Experiment Station, Cornell University, Geneva, New York, February 1999.

p. 91 *"In the transformation process:* This remark and the following material on Pioneer's gene gun is based on an author interview with Sam Wise.

p. 91 *One early marker researchers used:* Stephen Nottingham, *Eat Your Genes: How Genetically Modified Food Is Entering Our Diet* (New York: Zed Books, 1998), p. 23.

p. 95 *Some strains of tuberculosis:* See "Background on Antibiotic Resistance," Centers for Disease Control and Prevention, http://www.cdc.gov.

p. 95 *Advisers to the UK Ministry of Agriculture warned:* Marie Woolf, "Modified Corn on Sale in UK 'Kills' Life-Saving Antibiotics," *UK Independent,* June 6, 1999.

p. 95 *The FDA endorsed Calgene's use of a gene:* James Maryanski interviewed by John Henkel, *FDA Consumer* (FDA in-house publication), April 1995.

p. 96 *"See, the trouble with the whole industry:* Author interview with Sanford Miller.

p. 96 *Scientists can use markers:* One new marker gene comes from the jellyfish *Aequorea victoria.* Called green fluorescent protein, or GFP, the jellyfish gene can be inserted into plants along with the gene for the trait of interest. Plants containing the jellyfish gene glow green when an ultraviolet light is shined on them. See Brian Harper, Stephen Mabon, et al., "Green Fluorescent Protein as a Market for Expression of a Second Gene in Transgenic Plants," *Nature Biotechnology,* vol. 17, no. 11 (November 1999), pp. 1125–29. C. Neil Stewart, a researcher at the University of North Carolina at Greensboro and collaborator on this paper, described the use of this marker during remarks at a National Academy of Sciences meeting, July 13–14, 2000.

p. 98 *"No cell will ever make use of all:* Remarks by Ricarda Steinbrecher are based on an author interview and material from "Gene Files" by Ricarda Steinbrecher, Women's Environmental Network Trust, London.

p. 100 *To find out how widespread:* Jean Finnegan and David McElroy, "Transgene Inactivation: Plants Fight Back!" *Bio/Technology,* September 1994.

p. 100 *"As near as anyone can tell:* James Siedow, author interview. Siedow noted that part of the mechanism of gene silencing actually "has to do with very high levels of not the gene, but of the messenger RNA getting reproduced. That's one of the things that the plants can sense."

SIX   **PESTICIDE IN A SPUD**

p. 102 *It was first discovered in 1901:* "Beyond Discovery: The Path from Research to Human Benefit," National Academy of Sciences, October 1998.

p. 102 *In 1981 Helen Whitely and Ernest Schnepf:* Ibid.

p. 103 *"The crystals dissolve in the insect's gut:* Susan MacIntosh, author interviews.

p. 104 *In 1988 Monsanto requested permission: Genetically Modified Pest-Protected Plants: Science and Regulation,* National Academy of Sciences, 2000, p. 29.

p. 104 *While all organisms use the same basic genetic code:* On April 25, 1953, James D. Watson and Francis H. Crick published a short article entitled "A Structure for Deoxyribose Nucleic Acid" in the British journal *Nature.* They showed that DNA molecules carry the hereditary information for all forms of life. By the mid-1960s scientists had worked out the details of the genetic code by which four simple base pairs of deoxyribonu-cleic acid—adenine (A) and thymine (T), and cytosine (C) and guanine (G)—instruct the cells of all organisms, from the lowly krill to the baleen whale, to make twenty amino acids. The amino acids, in turn, build long chains of proteins. Proteins are the basic materials of cells. Some proteins are structural, making up bones, hair, skin, blood, leaves, petals, and other tissues. Others make pigments or hormones or enzymes. The totality of an organism's genetic information is called its genome. All the DNA of a plant or animal is carefully folded and tightly packed into chromosomes. Each species has a set number of chromosomes; for instance, there are forty-eight chromosomes in each cell of the chimpanzee, forty-six in humans, sixty-four in horses, twenty in corn, and forty-eight in potato plants.

p. 105 *As Ernest Jaworski told me, "It took:* Ernest Jaworski, author interviews.

p. 105 *"DNA and videotape are linear information:* Michael Voiland and Linda McCandless, "Development of the 'Gene Gun' at Cornell," Communications Services news, New York State Agricultural Experiment Station, Cornell University, Geneva, New York, February 1999; http://www.nyaes.cornell.edu/pubs/1999/genegun/html.

p. 106 *In 1990 Monsanto created the first successful:* See "Genetically Modified Pest-Protected Plants: Science and Regulation," National Academy of Sciences, p. 28, 2000; also, F. J. Perlak et al., "Insect Resistant Cotton Plants," *Bio/Technology,* vol. 8 (1990), pp. 939–43.

p. 106 Ernest Jaworski retired from Monsanto in 1991, before the company began its push to commercialize Bt crops.

p. 106 *Monsanto submitted a petition to the USDA:* "Petition for Determination of Nonregulated Status for Potatoes Producing the Colorado Potato Beetle Control Protein of *Bacillus thuringiensis* subsp. *tenebrionis,*" submitted by Terry B. Stone, Regulatory Affairs, Agricultural Group, Monsanto, to Michael A. Lidsky, Deputy Director, BBEP, APHIS, USDA, on September 7, 1994.

p. 108 *documents Monsanto submitted to prove:* For a summary of these documents ("Molecular Characterization of CPB Resistant Russet Burbank Potatoes Equivalence of Microbially-Produced Btt Protein; Equivalence of Microbially-Produced and Plant-Produced Btt Protein, etc."), see EPA, *Biopesticide Fact Sheet: Bacillus Thuringiensis CryIII(A) Delta Endotoxin and the Genetic Material Necessary for Its Production in Potato* (006432), EPA publication no. 730-F-00-008, April 2000, online at http://www.epa.gov/pesticides/biopesticides/factsheets/fs006432t.htm.

p. 109 *NAS argued that tests "should preferably:* Genetically Modified Pest-Protected Plants: Science and Regulation, NAS, 2000, pp. 65–66.

p. 109 *"The E. coli product is different:* Arpad Pusztai, author interview.

p. 109 *The gene they transplanted:* Arpad Pusztai et al., "Expression of the Insecticidal Bean α-Amylase Inhibitor Transgene Has Minimal Detrimental Effect on the Nutritional Value of Peas Fed to Rats at 30% of the Diet," *Journal of Nutrition,* vol. 129 (1999), pp. 1597–1603.

p. 112 *She has raised a concern that the naked:* Ho has written several articles detailing this thesis. See Mae-Wan Ho, A. Ryan, et al., "Cauliflower Mosaic Viral Promoter—A Recipe for Disaster?" *Microbial Ecology in Health and Disease,* vol. 11 (1999), pp. 194–97. Ho's book, *Genetic Engineering Dream or Nightmare? Turning the Tide on the Brave New World of Bad Science and Big Business,* 2d ed. (New York: Gateway, Gill & Macmillan, 1999), popularizes this concern.

p. 112 *"The vast majority of what she says:* James Siedow, author interview. Siedow is not alone in harshly criticizing Ho's concerns. See John Hodgson, "Scientists Avert New GMO Crisis," *Nature Biotechnology,* vol. 18, no. 1 (January 2000). Following an announcement from the journal *Microbial Ecology in Health and Disease* of Ho's forthcoming paper, Hodgson writes, "prominent plant scientists circulated their comments and criticisms rapidly on e-mail networks. The plant researchers uniformly damned the paper." Hodgson continues, "With considerable irony, several critics of the paper suggested that Ho et al. had not gone far enough in calling for a ban on transgenic crops containing the 35S promoter. 'Let's stop eating plants and animals altogether,' said Barbara Hohn. 'It's a shame we did not have this information millions of years ago. It would have been so easy to avoid the perils of life.' " Hodgson notes that the editor of *Microbial Ecology in Health and Disease,* Tore Midtvedt, a professor at the Karolinska Institute for the past eighteen years, "was aware the article would be controversial" and said his motivation was "to stimulate debate.' "

p. 114 *"Some results showed differences:* Royal Society, "Review of Data on Possible Toxicity of GM Potatoes," http://www.royalsoc.ac.uk/st_pol54.htm.

p. 115 *"The Monsanto B.t.-potato presents little:* See "Final Report of the FIFRA Scientific Advisory Panel Subpanel on Plant Pesticides, Meeting Held March 1, 1995," submittal from Robert B. Jaeger, designated federal official, to Daniel M. Barolo, director of the Office of Pesticide Programs.

p. 116 *They invited some of the world's top:* "Conference of Scientific

Issues Related to Potential Allergenicity in Transgenic Food Crops," April
18–19, 1994, transcript of proceedings, FDA docket no. 94N-0053 (obtained under the Freedom of Information Act).

SEVEN SOUND SCIENCE, STERILE SEEDS

p. 118 *"The benefits of agricultural biotechnology:* For opening state-
ments by Representative Ewing and other congressmen, as well as testi-
mony, and answers to questions posed by congressmen, given by USDA
Undersecretary August Schumacher, Roger Pine of the National Corn
Growers Association, Michael Yost of the American Soybean Association,
Tim Galvin, administrator with the USDA's Foreign Agricultural Service,
Assistant U.S. trade representative for agricultural affairs Jim Murphy,
Richard Parry, USDA scientist, and other witnesses, see House Committee
on Agriculture, Subcommittee on Risk Management, Research, and Spe-
cialty Crops, hearing on agricultural biotechnology, March 3, 1999, http://
commdocs.house.gov/committees/ag.

p. 119 *In 1998 U.S. farmers lost about $200 million worth of corn:*
USDA Undersecretary August Schumacher, testimony before the Senate Finance Subcommittee on International Trade, March 15, 1999.

p. 122 *The European Union refused to import beef:* The three natural
hormones approved by the U.S. government as safe are estradiol, proges-
terone, and testosterone; the three synthetic hormones are zeranol, tren-
bolone acetate, and melengestrol acetate.

p. 126 *"From the European perspective, it's almost:* Alan Simpson
remarks, press briefing sponsored by the Consumers Choice Council, Washington, D.C., February 25, 2000.

p. 126 *Pusztai's potatoes again exploded on the front pages:* Nigel
Poole, an official with AstraZeneca in the UK, provided samples of British
newspaper headlines he clipped between Feb. 5 and Feb. 25, 1999, as well
as statistics on column inches, during a presentation at the annual meeting
of the Biotechnology Industry Organization in Seattle, May 16–20, 1999.
(Poole also presented similar information at the International Business
Communications (IBC's) "Ag-Biotech World Forum," in Las Vegas, June
9–11, 1999.)

p. 127 *Two days later, on February 18, Greenpeace activists:* "Greenpeace Dumps 4 Tons of GE Soya at U.K. Prime Minister Tony Blair's Office," Reuters, February 18, 1999.

p. 127 *Blair published an article:* See *Daily Telegraph,* February 20,

1999. For a detailed account and timeline of the events surrounding the fallout from Pusztai's studies in the UK, see "Anatomy of a Food Scare," *New Scientist,* online at http://www.newscientist.com. See also Nuffield Council on Bioethics, "Genetically Modified Crops: The Ethical and Social Issues," Appendix 1: "Rats and Potatoes at the Rowett Institute," online at http://www.nuffieldfoundation.org.

p. 128 *Seymour, eighty-four years old, told:* John Seymour is quoted on the Voice of Irish Concern for the Environment website, Dublin, http://www.voice.buz.org.

p. 128 *The judge said he believed:* Kevin O'Sullivan, "Probation Act Applied to Six GM Food Protesters," *Irish Times,* April 1, 1999.

p. 129 *"The BMA believes that more research:* British Medical Association, Board of Science and Education, "The Impact of Genetic Modification on Agriculture, Food and Health, An Interim Statement," London, May 1999.

p. 129 *"Within a few years, virtually 100 percent:* Statement by Stuart Eizenstat, undersecretary for Economic, Business, and Agricultural Affairs, U.S. Department of State, before the Senate Finance Committee, Subcommittee on International Trade, March 15, 1999.

p. 130 *"We're looking forward very soon:* Remarks by Carl Feldbaum, Roger Beachy, William White, Steve Benoit, and Channapatna Prakash were delivered at the annual Biotechnology Industry Organization meeting in Seattle, May 16–20, 1999.

p. 133 *"We have a little catching up . . . :* Remarks by Brian Tokar, Martha Crouch, and Edward Hammond were presented at the Biodevastation3 meeting in Seattle, May 19–20, 1999. The first Biodevastation conference was held in St. Louis; Biodevastation2 was held in New Delhi.

p. 134 *The acquisition would give Monsanto:* Monsanto later withdrew from the proposed acquisition. See Monsanto Company, "Filing for Approval of Merger Between Monsanto and Delta & Pine Land Cancelled in Light of Regulatory Issues," news release, December 20, 1999.

p. 136 *Hundreds of Indian farmers:* Geeta Bharathan, "*Bt*-cotton in India: Anatomy of a Controversy," *Current Science,* vol. 79, no. 8 (October 25, 2000), pp. 1067–75.

p. 137 *Above the stunning picture, the headline:* Carol Kaesuk Yoon, "Altered Corn May Imperil Butterfly, Researchers Say," *New York Times,* May 20, 1999.

EIGHT **LETHAL CORN POLLEN**

p. 138 *"There's a lot. It's a very:* This and all ensuing remarks by John Losey are based on an author interview.

p. 142 *Mellon and her colleague:* "Transgenes for insecticidal or fungicidal compounds introduced into crop plants to inhibit pests, may kill non-target, and even beneficial, insects and fungi," they wrote prophetically in *The Ecological Risks of Engineered Crops* (Cambridge, Mass.: MIT Press, Massachusetts, 1996), p. 42.

p. 143 *"We're hearing from farmers:* Kathleen Hart, "Biotech Industry Backs Bt Corn, Questions Relevance of Report Showing Pollen Kills Monarch Butterflies," *Pesticide & Toxic Chemical News,* May 27, 1999.

p. 143 *"highly likely that in the natural:* For remarks by Val Giddings, Chris DiFonzo, and Warren Stevens, see Biotechnology Industry Organization, "Academic Researchers and Industry Agree Reports on Bt Crop Impact on Monarch Butterflies Overblown," press release, June 10, 1999.

p. 145 *"A review of existing . . . :* L. L. Wolfenbarger and P. R. Phifer, "The Ecological Risks and Benefits of Genetically Engineered Plants," *Science,* vol. 290 (December 15, 2000), p. 2088.

p. 145 *BIO's Val Giddings also asserted:* BIO press release, ibid.

p. 147 *"We must continue to argue in multilateral:* Kathleen Hart, "Organic Farming Booming, Fueled by Consumer Demand," *Pesticide & Toxic Chemical News,* March 25, 1999.

p. 148 *The United States "can't force these new:* Kathleen Hart, "Glickman Warns Agribusiness Against Putting All Its Eggs in Biotech Basket," *Pesticide & Toxic Chemical News,* May 6, 1999.

p. 149 *"If USDA says they're safe:* A corn and soybean grower in Garden City, Iowa, author interview.

p. 149 *The alliance, which one GMA official:* Information on the GMA's Alliance for Better Foods is based on an author interview with GMA vice president Gene Grabowski.

p. 151 *"Following years of testing . . . :* "Testimony of Matthew G. Stanard," GMA, http://www.gmabrands.com.

p. 151 *once a society agrees that "process" has validity:* Kathleen Hart, "Monsanto Changes Stand on Labeling Genetically Modified Food in European Union," *Pesticide & Toxic Chemical News,* May 7, 1998.

p. 153 *During that time Monsanto:* Peter Montague, "Hormones in

Milk: No Right to Know," *Rachel's Hazardous Waste News* 381 (March 17, 1994).

p. 153 *Monsanto announced that Taylor would:* Monsanto, "Michael R. Taylor Named Vice President for Public Policy at Monsanto," press release, September 10, 1998.

p. 153 *Monsanto culled dozens of lobbyists:* For a detailed listing of Monsanto lobbyists and congressional contributions, see Center for Public Integrity, *Unreasonable Risk: The Politics of Pesticides* (Washington, D.C.: Center for Public Integrity, 1998).

p. 154 *"far exceeds our ability to understand:* Letter from Carl Pope to President Clinton, August 18, 1999.

p. 155 *Gerber officials in Michigan apparently:* Lucette Lagnado, "Gerber Baby Food, Gilled by Greenpeace, Plans Swift Overhaul," *Wall Street Journal,* July 30, 1999.

p. 155 *A Heinz official told:* "Gerber Drops Bioengineered Suppliers," Associated Press, August 1, 1999.

p. 156 *"No long-term studies of the impact:* For the comments by Martha Herbert, Paul Billings, and the professional societies, see: Kathleen Hart, "Consumer Groups Call on Clinton Administration to Perform Rigorous Research on Genetically Engineered Crops," *Pesticide & Toxic Chemical News,* August 26, 1999.

p. 157 *"At the same time, neither . . .":* Testimony of Mark Silbergeld before the Committee on Agriculture, U.S. Senate hearings on biotechnology, October 7, 1999.

p. 157 *"We recognize the policy is stale:* Kathleen Hart, "AMA to Revisit Policy on Genetically Engineered Crops," *Pesticide & Toxic Chemical News,* September 2, 1999.

p. 158 *At each listening session:* For public comments during the listening sessions in St. Paul and Burlington, see U.S. Department of Agriculture, Foreign Agricultural Service, online transcripts, http://www.fas.usda.gov/itp/wto.

NINE   **DYING ON THE VINE**

p. 161 *Joyce Nettleton, director:* Joyce Nettleton, the German counselor for agriculture, Tassos Haniotis, and the U.S. trade representative official gave presentations at the program "Genetically Engineered Organisms: International Considerations," sponsored by the Institute of Food Technologists, Washington, D.C., September 13, 1999.

p. 163 *Agricultural biotechnology, "which holds so much:* Representative Nick Smith and Anthony Shelton, remarks during House of Representatives, Committee on Science, Subcommittee on Basic Research, hearing on "Plant Genome Science: From the Lab to the Field to the Market, Part II," Washington, D.C., October 5, 1999.

p. 163 *They confirmed that Bt corn:* Entomological Society of America, "Study Confirms Gene-Altered Corn Kills Monarch Butterflies," news release, November 3, 1999.

p. 164 *"We found this data that there could be:* John Losey, author interview.

p. 165 *"The biotech ball game:* Senator Richard Lugar, Dean DellaPenna, and Gary Kushner, remarks during Senate Committee on Agriculture, Nutrition, and Forestry, hearing on biotechnology, October 7, 1999.

p. 166 *DellaPenna had acknowledged elsewhere:* Dean DellaPenna, "Nutritional Genomics: Manipulating Plant Micronutrients to Improve Human Health," *Science,* vol. 285, no. 5426 (July 16, 1999), pp. 375–79.

p. 167 *"I think we've tended to see:* This and ensuing comments by Robert Shapiro were presented during video broadcast by Monsanto; see also Robert Shapiro, address to Greenpeace Business Conference, London, October 6, 1999, transcript by HearingRoom.com. See Kathleen Hart, "Rockefeller Foundation Advocates Ag Biotechnology for 'Doubly Green Revolution,' " *Pesticide & Toxic Chemical News,* November 18, 1999.

p. 168 *Conway believes that food plants:* Gordon Conway, remarks at "Agbiotech99: Biotechnology and World Agriculture" conference, sponsored by *Nature Biotechnology,* London, November 14–16, 1999.

p. 168 *"I am writing to let you know . . . :* Letter from Robert Shapiro, Chairman and CEO of Monsanto, to Gordon Conway, President, Rockefeller Foundation, October 4, 1999.

p. 170 *"Don't look, don't find":* Margaret Mellon, author interview.

p. 171 *The academy found that many:* For a detailed account of the problems with Monsanto's studies, see National Research Council, *Genetically Modified Pest-Protected Plants: Science and Regulation* (Washington, D.C.: National Academy Press, 2000).

p. 172 *The companies set up a $100,000 kitty:* Kathleen Hart, "EPA Looking to Industry Study to Guide Agency Decisions on Ecological Safety of Bt Corn," *Pesticide & Toxic Chemical News,* September 30, 1999.

p. 173 *"What the group hears today:* Janet Andersen, Orley R. "Chip" Taylor, Robert Hartzler, Doug Buhler, Mark Sears, Galen Dively, and Richard Hellmich, remarks to "Monarch Butterfly Research Symposium," Rosemont, Ill., November 2, 1999.

p. 176 *Midway through the afternoon:* Agricultural Biotechnology

Stewardship Working Group, "Scientific Symposium to Show No Harm to Monarch Butterfly," media advisory, November 2, 1999.

TEN GLOBAL FOOD FIGHT

p. 178 *What took place:* The material in this chapter is based on presentations, press conferences, interviews, literature compiled and distributed by delegates, and coverage both of official and NGO venues at the World Trade Organization meeting in Seattle from November 30 to December 4, 1999. Also see Kathleen Hart, "Eyewitness Account: Melee in Seattle Traps Reporter, Delegates," "Clinton Administration Seeks Greater Market Access for U.S. Agricultural Products," and "Seattle Protests Likely to Prompt WTO Policy, Operational Changes," *Food Chemical News,* December 6, 1999; and Kathleen Hart, "WTO Ministers Fail to Agree on Environment, Agriculture, Biotech Issues," *Pesticide & Toxic Chemical News,* December 9, 1999.

p. 182 *"on the enhancement of the power . . .":* This and other quotes from Ralph Nader, preface to *Whose Trade Organization? Corporate Globalization and the Erosion of Democracy,* by Lori Wallach and Michelle Sforza (Washington, D.C.: Public Citizen Foundation, 1999), ix.

p. 186 *"Present scientific knowledge:* Remarks by Benedikt Haerlin presented at "Agbiotech99: Biotechnology and World Agriculture," conference sponsored by *Nature Biotechnology,* London, November 14–16, 1999.

p. 192 *"While some would have you believe:* Yoko Tomiyama, chairperson of Consumers Union of Japan, "An Open Letter to American Farmers and Agribusiness," October 6, 1999.

p. 193 *This time the 132 countries:* This account of the Montreal negotiations is based on interviews with NGOs who were present for the Biosafety Protocol talks; also, an environmental NGO perspective by Mark S. Winfield, "Reflections on the Biosafety Protocol Negotiations in Montreal," Canadian Institute for Environmental Law and Policy, posted online at: www.biotech-info.net/BSP_reflections.html

ELEVEN SEEDS OF DISPUTE

p. 194 *On December 13, 1999, more than a thousand:* Jessie Seyfer, "Protesters Rally vs. 'Frankenfoods,'" Associated Press, December 13, 1999.

p. 195 *"Any risks from genetically engineered:* Susanne Huttner, Ignacio Chapela, Jo Ann Baumgartner, Nell Newman, Melodi Nelson, Robert Cannard, Roy Fuchs, Kara Cosby, Paul Bettencourt, Mike Phillips, and Janet Brown presented their views at the FDA public meeting "Biotechnology in the Year 2000 and Beyond," Oakland, California, December 13, 1999.

p. 199 *"I've never raised GMO crops:* Naylor spoke at a press briefing sponsored by the Consumers Choice Council in Washington, D.C., February 25, 2000.

p. 201 *George Naylor and his wife, Peggy, joined: Bruce Pickett, George Higginbotham, George Naylor, Peggy Naylor and Patrick de Kochko v. Monsanto Company,* class-action complaint filed December 14, 1999, in the U.S. District Court for the District of Columbia.

p. 201 *"It is a concern of every farmer:* Elizabeth Cronise spoke at the Consumers Choice Council press briefing in Washington, D.C., February 25, 2000.

p. 202 *In 1998 Monsanto charged the Saskatchewan farmer:* See *Monsanto Canada Inc. and Monsanto Company v. Percy Schmeiser and Schmeiser Enterprises Ltd.,* Federal Court of Canada, Docket T-1593-98.

p. 204 *"It may soon become impossible . . . :* National Farmers Union press release, Saskatchewan, "Biotech Companies Must Pay for Genetic Pollution," May 10, 1999.

p. 204 *"Farmers have been caught:* Remarks by Gary Goldberg presented at a meeting sponsored by the Worldwatch Institute in Washington, D.C., February 14, 2000.

p. 206 *"After being presented with a factual account:* FDA Center for Food Safety and Applied Nutrition, Office of Scientific Analysis and Support, "Report on Consumer Focus Groups on Biotechnology," October 20, 2000.

TWELVE GOLDEN RICE

p. 210 *At a science meeting sponsored:* Ingo Potrykus, "Vitamin-A and Iron-Enriched Rices May Hold Key to Combating Blindness and Malnutrition: A Biotechnology Advance," presented at "Agbiotech 99: Biotechnology and World Agriculture," conference sponsored by *Nature Biotechnology,* London, November 14–16, 1999.

p. 211 *"One of the things that we . . . :* Transcript: "Senator Bond, Roger Beachy Roundtable on Biotechnology," Bangkok, U.S. Department of State, January 21, 2000.

p. 213 *The week after Clinton's endorsement:* "This Rice Could Save a Million Kids a Year," *Time,* July 31, 2000, cover.

p. 213 *Greenpeace challenged the notion:* See "Genetically Engineered Pro-vitamin A 'Golden Rice' Reality vs. Fiction" and "The False Promise of Genetically Engineered Rice," Greenpeace website, http://www.greenpeace.org/~geneng.

p. 214 *According to the study, the "best:* See Xudong Ye, Salim Al-Babili, et al., "Engineering the Provitamin A (β-Carotene) Biosynthetic Pathway into (Carotenoid-Free) Rice Endosperm," *Science,* vol. 287, January 14, 2000, pp. 303–305.

p. 214 *Potrykus responded that nutritionists:* The rebuttal to Greenpeace was posted on the AgBioWorld Foundation website, http://www.agbioworld.org. Channapatna Prakash, a professor at Tuskegee University, is president of the foundation.

p. 215 *"You don't live by corn:* Vandana Shiva, Channapatna Prakash, and Dennis Kucinich were among the speakers at the Capitol Hill debate "Can Biotechnology Help Fight World Hunger?" June 29, 2000.

p. 217 *Egziabher said he was "appalled:* Letter from Tewolde Berhan Gebre Egziabher to Channel 4 Television and *The Times* (London), in protest to the documentary *Equinox* (aired March 19, 2000) and the article "GM Foods and the Luxury of Choice" (March 21, 2000), using southern scientists to make Europeans feel guilty for not supporting genetic engineering.

p. 217 *"Private sector multinationals . . . :* Florence Wambugu, *Modifying Africa,* available from the website: http://www.modifyingafrica.com.

p. 218 *The United Nations Development Program:* Sakiko Fakuda-Parr, *Human Development Report 2001: Making New Technologies Work for Human Development,* http://www.undp.org/hdr2001/.

p. 218 *Canola experts said it was the first:* Mary MacArthur, "Triple-Resistant Canola Weeds Found in Alta," *Western Producer,* February 10, 2000.

p. 219 *A German scientist found that a gene:* Antony Barnett, "GM Genes 'Jump Species Barrier,' " *Observer/Guardian,* May 28, 2000, online at http://www.guardian.co.uk/Archive/.

p. 219 *scientists had just discovered two "unexpected":* "Re: Roundup Ready Soybean Even 40-3-2, Petition 93-258-01P," letter from Monsanto's Washington, D.C., office to a USDA official in Riverdale, Maryland, May 15, 2000. New studies showed that the genetically engineered soybean contained "an additional, previously unobserved 250 bp segment of the

CP4 EPSPS element on the primary functional insert and a second insert consisting of 72 bp of the CP4 EPSPS sequence," Monsanto's director of regulatory affairs wrote. "In addition, these recent studies show that the NOS transcriptional termination sequence is intact, and not a partial element as previously reported." A year later, in May 2001, Pieter Windels and colleagues reported finding an unknown DNA segment in Monsanto's genetically engineered soybeans. See Pieter Windels, Isabel Taverniers, et al., "Characterisation of the Roundup Ready Soybean Insert," *European Food Research and Technology,* vol. 213, issue 2 (2001), pp. 107–12; online at http://link.springer.de/search.htm.

p. 220 *Vencill and his colleagues found that the stems:* The bacterial enzyme that confers resistance to the compound glyphosate "affects a major metabolic pathway in the plant, and has the side effect of sending lignin production 'into overdrive,' " Vencill said. See Andy Coghlan, "Splitting Headache: Monsanto's Modified Soya Beans Are Cracking Up in the Heat," *New Scientist,* November 20, 1999.

p. 220 *In the spring of 1999, Marc Lappé:* See Marc A. Lappé, E. Britt Bailey, et al., "Alterations in Clinically Important Phytoestrogens in Genetically Modified, Herbicide-Tolerant Soybeans," *Journal of Medicinal Food,* vol. 1, no. 4 (July 1999).

p. 221 *The American Soybean Association:* For background information on "Soybean Composition," "Soybean Health Benefits," and "Roundup Ready Soybeans," see American Soybean Association website http://www.oilseeds.org/asa/documents.

p. 221 *Opponents of genetically engineered:* See Tom Abate, "Coke Scion Joins Protest Against Genetically Engineered Ingredients," *San Francisco Chronicle,* April 3, 2000.

p. 222 *"Why rush unproven products:* See Ariane van Buren, "Genetically Engineered Food: Rush to Ruin?" *USA Today,* February 2000; press release, Interfaith Center on Corporate Responsibility.

p. 222 *At Quaker Oats Company:* For outcomes of shareholder initiatives, see "Quaker Shareholders Reject Ban on GMO Ingredients," Reuters, May 10, 2000; and "Shareholders Educate Management on Genetically Engineered Products," September 21, 2000, Socialfunds.com.

p. 223 *Frito-Lay told the Associated Press:* David Koenig, "Frito-Lay Angers Farm Groups," Associated Press, February 1, 2001.

p. 223 *consumer research "shows that . . . :* Monsanto statement, April 28, 2000, Monsanto Media Center, http://www.monsanto.com/monsanto/mediacenter/background/00apr28_newleaf.html.

p. 223 *"The real question in front of us . . .* : Remarks by Charles Margulis and Peter Hoffman, GE Food Alert news briefing, Washington, D.C., July 19, 2000.

THIRTEEN STARLINK AND TACOS

p. 225 *Close to midnight on July 25, 2000:* Information on the discovery of StarLink corn contamination is based on author interviews with Larry Bohlen.

p. 226 *polymerase chain reaction, or PCR:* This lab procedure allows biochemists to scan a sample of DNA for particular strings of DNA, called target sequences. Laboratory technicians start with primers, or small DNA molecules with sequences that correspond to the target sequence. A heat-stable enzyme is then added to the sample, and new copies of the target sequence are synthesized. From a single DNA molecule, billions of copies can be generated within a few hours.

p. 226 *Andrew Cockburn, an Aventis official, told the EPA:* Andrew Cockburn, "Aventis Safety Assessment of Cry9C Insecticidal Protein," submitted to EPA Scientific Advisory Panel, February 29, 2000.

p. 227 *"When only some of the corn is:* Kathleen Hart, "Biotech Company Tells EPA Novel Protein Poses no Risk for Food Allergies," *Pesticide & Toxic Chemical News,* March 9, 2000.

p. 227 *"Based on the available data:* SAP Report No. 2000-01A, "Food Allergencity of Cry9C Endotoxin and Other Non-digestible Proteins," June 29, 2000, EPA.

p. 230 *"The first reaction we had was:* Dan Lynn of Azteca Milling, remarks at the StarLink Summit sponsored by the Hudson Institute, Washington, D.C., December 6, 2000.

p. 231 *Kraft Foods announced a voluntary recall:* "Kraft Foods Announces Voluntary Recall of All Taco Bell Taco Shell Products from Grocery Stores," September 22, 2000, Special Report, online at http://www.kraft.com/special_report.

p. 231 *Sales from the taco line of foods account:* Scott Kilman and Sarah Lueck, "Kraft Taco Shell Puts Focus on Biotechnology Oversight," *Wall Street Journal,* September 25, 2000.

p. 231 *"We believe that this situation:* Grocery Manufacturers of America, press release, September 22, 2000.

p. 232 *"At stake, is the reputation and integrity:* National Corn Growers Association, press release, September 23, 2000.

p. 232 *"The challenge here is:* Comments by Senators Jim Jeffords, Tom Harkin, and Barbara Boxer, as well as Joe Levitt, Senate hearing, September 26, 2000.

p. 233 *Quaker Oats Company would require suppliers:* Susan Kelly, "Food Firms Scour Supply Chain After Kraft Recall," Reuters, September 26, 2000.

p. 234 *Aventis CropScience suspended sales of StarLink:* "Aventis Crop-Science Taking Immediate Action to Assure Confidence in StarLink Corn Distribution," Aventis press release, September 26, 2000.

p. 234 *As Aventis and the USDA began tracking down farmers:* Advisory Committee on Agricultural Biotechnology meeting, Washington, D.C., April 18, 2001.

p. 235 *Ralph Klemme, an Iowa legislator,:* K. T. Arasu, "Iowa Legislator-Farmer Cries Foul over StarLink," Reuters, November 7, 2000.

p. 235 *The extent of adulteration of the food supply:* See FDA Enforcement Report, "Class II Recalls" November 1, 2000, and "Class II Recalls," November 15, 2000.

p. 236 *a major breakfast cereal company shut down:* "Kellogg Reportedly Shut Plant Over Genetically Altered Corn," *New York Times,* October 22, 2000. While the *Times* reported that Kellogg officials "would not confirm the shutdown" to a *Washington Post* reporter, food company officials later alluded to the incident during a meeting of the Advisory Committee on Agricultural Biotechnology in Washington, D.C., April 18, 2001.

p. 236 *the AP reported, supplies of Cheetos:* "Testing Corn Affects Cheetos Supply," Associated Press, December 9, 2000.

p. 236 *DNA testing is expensive:* Susan Harlander, comments to EPA Scientific Advisory Panel Meeting, EPA, Rosslyn, Va., November 28, 2000.

p. 237 *Japan, the largest buyer, typically purchases:* In 1998 corn exports to Japan represented $1.5 billion of the total $5 billion in revenue from U.S. corn exports, according to the USDA.

p. 239 *In a review of the Aventis petition:* See "EPA Preliminary Evaluation of Information Contained in the October 25, 2000 Submission From Aventis Crop Science."

p. 240 *"Aventis does not know how:* Press release, "Aventis Crop-Science Finds Bioengineered Protein in Non-StarLink Corn Seed," FDA website, http://www.fda.gov/oc/po/firmrecalls/aventis11_00.html.

p. 244 *In November 1998 the Georgetown University Center:* Grocery Manufacturers of America, "GMA Forms Strategic Alliance with Georgetown University Center for Food and Nutrition Policy," news release, November 5, 1998.

p. 244 *Since the early 1990s, Hoban has been conducting:* See Thomas
Hoban and Lisa Katic, "American Consumer Views on Biotechnology,"
Cereal Foods World, vol. 43, no. 1 (1998), pp. 20–22, and Thomas J.
Hoban, Eric Woodrum, et al., "Public Opposition to Genetic Engineer-
ing," *Rural Sociology,* vol. 57, no. 4 (1992), pp. 476–93.

p. 244 *"Americans remain positive over the benefits:* GMA, "GMA
Survey Shows Americans Learning More About Biotechnology; Food Consumption Patterns Unchanged," news release, October 12, 2000.

p. 245 *Nearly 60 percent of women expressed concern:* "Many Americans Say Stop Planting Gene-Altered Crops," Reuters, November 3, 2000.

p. 245 *In the Pulse survey, only 50 percent of American women:* Oxy-
gen/Markle Pulse study conducted November 11, 2000, "Pulse Study:
Americans Uncertain about Genetically Modified Foods," online at http://
pulse.oxygen.com.

p. 246 *"Many observers fear that this could be the event: Jurassic
Foods? The Food Industry in a Post-StarLink World,* Promar International,
press release, Alexandria, Virginia, October 24, 2000, http://www.promar-
international.com/food_and_fear.html.

p. 246 *"The StarLink incident hasn't changed public opinion:* Lester
Crawford, author interview.

p. 246 *The EPA asked the panel:* Remarks by Carol Rubin, Michael
Phillips, Margaret Wittenberg, Larry Bohlen, Hugh Sampson, Marc
Rothenberg, delivered at the EPA's Scientific Advisory Panel meeting, Ross-
lyn, Va., November 28, 2000.

p. 249 *"Due to the mysterious and sudden:* Letter from Sharon Wolff
to Carol Browner, "Re: Docket No. PF-867B & Docket No. OPP-00688,"
November 25, 2000, Public Information and Records Integrity Branch,
Office of Pesticide Programs, EPA.

p. 250 *People can have IgE allergic reactions to pollen:* Food allergies
actually account for only a small portion of allergic diseases.

p. 251 *"Today we are in a very different:* Remarks by Janet Andersen,
Dean Metcalfe, Marc Rothenberg, Susan MacIntosh, Keith Finger, Grace
Booth, and Hugh Sampson, delivered at EPA's Scientific Advisory Panel
meeting, July 17–18, 2001.

p. 251 *Dean Metcalfe, an allergy specialist:* In particular, Metcalfe
voiced the concern at the meeting that "if processing changed the Cry9C
protein, it curled it differently or exposed different sequences or put different sequences together, a person reacting could potentially react to those
epitopes which would not be seen with the standards that are used."

FOURTEEN **FUTURE FOOD SECURITY**

p. 255 *"It means that the old paradigm that one gene:* Barbara Culli-ton, "One Gene, Many Proteins," February 12, 2001, online at http://www.celera.com.

p. 256 *"Genetic determinism is dead,":* Mae-Wan Ho, "Human Gen-ome Map Spells the Death of Genetic Determinism," *ISIS News* [Institute of Science in Society], February 2001.

p. 256 *"From a scientific perspective, the public argument:* Robert Goodman is quoted in a University of Wisconsin–Madison news release, February 18, 2001. Additional comments by Goodman are based on au-thor interview.

p. 258 *"Recombinant DNA molecular research:* Donald N. Duvick, "Recombinant DNA Molecule Research Potential for Agricultural Crop Plants," presentation to the National Research Council, April 13, 1977; online at http://www.biotech-info.net/recombinant_DNA_Duvick3.html.

p. 258 *"Most of the promises that are being offered:* Wes Jackson, re-marks during a debate on benefits and risks of biotechnology sponsored by Environmental Media Services, Washington, D.C., January 24, 2000.

p. 259 *"Some of the world's leading authorities:* Alan Simpson, re-marks at Consumers Choice Council press briefing, Washington, D.C., February 25, 2000.

p. 259 *"The products in commercial production:* F. L. Erickson and P. G. Lemaux, "Issues Related to the Development and Use of Engineered Herbicide-Tolerant Crops in California," presentation at the California Weed Science Society conference in Sacramento, Calif., January 10, 2000.

p. 260 *"Our method is ready for use by plant biologists:* Peggy Lemaux is quoted in a University of California–Berkeley press release, "Jumping Genes Ease Perceived Safety Concerns, Promote Stability in Genetically Modified Cereals, UC Berkeley Researcher Finds," March 6, 2001.

p. 260 *Peggy Lemaux and her colleague Thomas Koprek:* See Thomas Koprek et al., "Transposon-Mediated Single-Copy Gene Delivery Leads to Increased Transgene Expression Stability in Barley," *Plant Physiology,* vol. 125 (March 2001), pp. 1354–62.

p. 261 *"We are successfully integrating:* Hendrik Verfaillie, letter to shareowners, March 1, 2001, online at http://www.monsanto.com. For Roundup Sales, see http://monsanto.com/monsanto/media/01/01feb12_invest.html.

p. 262 *But in February 2001 the Royal Society of Canada:* Expert Panel Report on the Future of Food Biotechnology, prepared by the Royal Society of Canada at the request of Health Canada, Canada Food Inspection Agency and Environment Canada, February 5, 2001.

p. 262 *But he also believes "there were:* Brian Ellis, author interview.

p. 265 *Extensive ecological studies:* See *Proceedings of the National Academy of Sciences,* online at http://www.pnas.org/cgi/content/full/211297698v1.

p. 265 *United States "has a regulatory system:* Jane Rissler, author interview.

p. 266 *There is "effectively no federal standard:* Consumer Federation of America Foundation, "Report Says U.S. Regulation of Genetically Modified Foods Includes Huge Loopholes that Permit Marketing with Little Government Oversight," news release, January 11, 2001.

p. 266 *"Consumers should have the right:* EU–U.S. Biotechnology Consultative Forum Final Report, December 18, 2000.

p. 267 *Federal regulatory oversight:* Summary, "Report 10 of the Council on Scientific Affairs," American Medical Association, December 14, 2000.

p. 268 *"Unfortunately, we've no systematic way:* Elizabeth Cullen, author interview.

p. 269 *the FDA proposed formal regulations of bioengineered:* FDA, "FDA Announces Proposal and Draft Guidance for Food Developed Through Biotechnology," Premarket Notice Concerning Bioengineered Foods, a proposed rule, January 17, 2001; and "Guidance for Industry Voluntary Labeling Indicating Whether Foods Have or Have Not Been Developed Using Bioengineering." This rule was the first formal regulation to be proposed by FDA to regulate genetically engineered foods. Three months earlier, on September 29, 2000, the U.S. District Court for the District of Columbia dismissed the Alliance for Bio-Integrity's challenge (see Chapter 2) to the FDA's 1992 "Statement of Policy: Foods Derived From New Plant Varieties." Judge Colleen Kollar-Kotelly ruled that the FDA policy statement was not a formal regulation. In an October 6, 2000, "Talk Paper" released by the FDA on the ruling, the FDA stated: "The court agreed with FDA that the policy statement was not a rule requiring notice and comment rulemaking. . . . The court deferred to FDA's view that genetically engineered foods as a class do not require premarket review and approval of a food additive petition. The court also accepted FDA's view that special labeling for genetically engineered foods as a class is not required solely because of consumer demand or because of the process used to develop these foods. Finally, the court ruled that the plaintiffs' first amendment rights were not violated, nor does the 1992 policy violate the Religious Freedom Restoration Act."

p. 271 *By late October 2001:* Remarks by Linda Kahl delivered at a conference on the future of food biotechnology, sponsored by the Food and Drug Law Institute, Washington, D.C., October 29, 2001.

p. 271 *Organic farmers, conventional growers:* Comments of Edward Korwek, Michael Yost, Hunt Shipman, and Mary-Howell Martens were made during the USDA's Advisory Committee on Agricultural Biotechnology meeting, Washington, D.C., April 17–18, 2001.

p. 273 The Wall Street Journal *tested the extent:* Patricia Callahan and Scott Kilman, "Some Ingredients Are Genetically Modified, Despite Labels' Claims," *Wall Street Journal,* April 5, 2001.

p. 273 *"Genetically engineered crops:* e-mail communication from Craig Winters.

p. 274 *"Japanese consumers are highly suspicious:* "Japanese Millers State Opposition to GM Wheat-Group," Reuters, February 22, 2001.

p. 275 *"I think the reason why the original:* Judge Barth, author interview.

p. 276 *"Our flour's good for baking bread:* John Ellis, author interview.

p. 277 *President George W. Bush proclaimed:* President of the United States, National Biotechnology Month, January 2001, press release, January 20, 2001.

p. 278 *"It's been all pain and no gain:* Gene Grabowski, author interview.

p. 280 *In November 2001 Ignacio Chapela reported:* David Quist and Ignacio Chapela, "Transgenic DNA introgressed into traditional maize land races in Oaxaca, Mexico," *Nature,* vol. 414, November 29, 2001, pp. 541–43. See also University of California, Berkeley, press release, "Transgenic DNA Discovered in Native Mexican Corn, According to a New Study by UC Berkeley Researchers," November 28, 2001.

p. 280 *The first farmers gathered grain:* See "Primal Seeds," online at http://www.primalseeds.org/agricult.htm. For an account of the beginnings of wheat cultivation, see John Noble Wilford, "New Clues Show Where People Made the Great Leap to Agriculture," *New York Times,* November 18, 1997.

ACKNOWLEDGMENTS

Lilia Adecer Cajilog, a passionate defender of indigenous rights and native plants, reminded me as I began writing that a book's author is only its co-creator. I owe a deep debt of gratitude to Lilia and to:

Phil Wallace, a friend and colleague, who first urged me to undertake the writing of this book, discussed ideas with me throughout its evolution, and offered his unflagging encouragement;

Terry Tempest Williams, a naturalist and writer of amazing talents, who believed instantly in the rightness of this project and my ability to carry it to fruition;

My editors in New York, Dan Frank, who graciously provided me with the time and support I needed to write the book, and Andrew Miller, whose keen editorial insights helped to shape and sharpen its final form;

Jane Rissler, a scientist at the Union of Concerned Scientists, who opened her mind and library so generously to me and reviewed the first draft of the manuscript; and

Jay Fletcher, my editor at *Food Chemical News,* who supported my dogged pursuit of biotech food stories across the United States and the European Union for more than four years.

Many scientists, from diverse backgrounds, gave freely and patiently of their knowledge on subjects ranging from molecular biology to ecology, some during repeated conversations, others in a single key interview. In particular, I wish to thank Arpad Pusztai, Susan MacIntosh, John Losey, Brian Ellis, Michael Hansen, Ernest Jaworski, Robert Goodman, and James Siedow for their explanations of complex scientific material, while at the same time taking sole responsibility for any erroneous conclusions I may have drawn from their words. I also want to thank Aaron Bouchie at *Nature Biotechnology* and Frances McKim at AgriSolutions in the UK.

My deep appreciation goes to Juneanne Kimbrough and Amy Holmes, who believed ardently in the importance of telling Americans about genetically engineered foods and gave me boundless encouragement; to David Noble, who read the entire first draft and offered enthusiastic support; to Dee Redfearn and Lyssa Tall, who read early chapters and provided valuable perspective; to Lee Gutkind and other inspiring creative nonfiction writers at the Goucher College conference in August 1999; to Janet Biehl, copy editor, and Grace McVeigh, production editor at Pantheon Books; and to Phil Zahodiakin, Janet Byron, Clif Wiens, Bryan Lee, George Ovitt Jr., Dorothy Demmel, Michael Keith, Nancy Hart, Wilson Dizard, Consuelo Parson, Elizabeth Tsang, and Jack Connor.

Finally, for her love, understanding, and finely tuned perceptions, I thank my daughter, Alexis Hart Ovitt, who inspired me to explore what the genetic alteration of food might mean for her and her generation's children.

ABOUT THE AUTHOR

Kathleen Hart is a journalist who has been writing about health and the environment for fifteen years. She has covered agriculture and biotechnology for *Food Chemical News* and reported on nuclear energy and nonproliferation for McGraw-Hill's *Nucleonics Week*. She previously edited the *Environmental Health Letter*. Her articles have appeared in various publications, including the *Boston Globe* and the *Bulletin of the Atomic Scientists*. Hart lives in Washington, D.C.